228
Advances in Polymer Science

AF786136

Advances in Polymer Science

Recently Published and Forthcoming Volumes

Complex Macromolecular Systems II
Volume Editors: Müller, A.H.E., Schmidt, H.-W.
Vol. 228, 2010

Complex Macromolecular Systems I
Volume Editors: Müller, A.H.E., Schmidt, H.-W.
Vol. 227, 2010

Shape-Memory Polymers
Volume Editor: Lendlein, A.
Vol. 226, 2010

Polymer Libraries
Volume Editors: Meier, M.A.R., Webster, D.C.
Vol. 225, 2010

Polymer Membranes/Biomembranes
Volume Editors: Meier, W.P., Knoll, W.
Vol. 224, 2010

Organic Electronics
Volume Editors: Meller, G., Grasser, T.
Vol. 223, 2010

Inclusion Polymers
Volume Editor: Wenz, G.
Vol. 222, 2009

Advanced Computer Simulation Approaches for Soft Matter Sciences III
Volume Editors: Holm, C., Kremer, K.
Vol. 221, 2009

Self-Assembled Nanomaterials II
Nanotubes
Volume Editor: Shimizu, T.
Vol. 220, 2008

Self-Assembled Nanomaterials I
Nanofibers
Volume Editor: Shimizu, T.
Vol. 219, 2008

Interfacial Processes and Molecular Aggregation of Surfactants
Volume Editor: Narayanan, R.
Vol. 218, 2008

New Frontiers in Polymer Synthesis
Volume Editor: Kobayashi, S.
Vol. 217, 2008

Polymers for Fuel Cells II
Volume Editor: Scherer, G.G.
Vol. 216, 2008

Polymers for Fuel Cells I
Volume Editor: Scherer, G.G.
Vol. 215, 2008

Photoresponsive Polymers II
Volume Editors: Marder, S.R., Lee, K.-S.
Vol. 214, 2008

Photoresponsive Polymers I
Volume Editors: Marder, S.R., Lee, K.-S.
Vol. 213, 2008

Polyfluorenes
Volume Editors: Scherf, U., Neher, D.
Vol. 212, 2008

Chromatography for Sustainable Polymeric Materials
Renewable, Degradable and Recyclable
Volume Editors: Albertsson, A.-C., Hakkarainen, M.
Vol. 211, 2008

Wax Crystal Control · Nanocomposites Stimuli-Responsive Polymers
Vol. 210, 2008

Functional Materials and Biomaterials
Vol. 209, 2007

Phase-Separated Interpenetrating Polymer Networks
Authors: Lipatov, Y.S., Alekseeva, T.
Vol. 208, 2007

Hydrogen Bonded Polymers
Volume Editor: Binder, W.
Vol. 207, 2007

Oligomers · Polymer Composites Molecular Imprinting
Vol. 206, 2007

Polysaccharides II
Volume Editor: Klemm, D.
Vol. 205, 2006

Neodymium Based Ziegler Catalysts – Fundamental Chemistry
Volume Editor: Nuyken, O.
Vol. 204, 2006

Complex Macromolecular Systems II

Volume Editors: Axel H.E. Müller
Hans-Werner Schmidt

With contributions by

V. Altstädt · H. Audorff · M. Ballauff · A. Böker
D.P. Erhard · R. Giesa · D. Haarer · S. Huettner
L. Kador · K. Kreger · S. Kutuzov · D. Lovera
A.H.E. Müller · F. Plamper · N. Popp · H.-W. Schmidt
M. Sommer · M. Thelakkat · C. von Salis-Soglio
R. Walker · Y. Xu

Springer

Editors
Prof. Axel H.E. Müller
Universität Bayreuth
Lehrstuhl für Makromolekulare Chemie II
Universitätsstr. 30
95440 Bayreuth
Germany
axel.mueller@uni-bayreuth.de

Prof. Hans-Werner Schmidt
Universität Bayreuth
Lehrstuhl für Makromolekulare Chemie I
Universitätsstr. 30
95440 Bayreuth
Germany
hans-werner.schmidt@uni-bayreuth.de

ISSN 0065-3195 e-ISSN 1436-5030
ISBN 978-3-642-26344-6 ISBN 978-3-642-12912-4(eBook)
DOI 10.1007/978-3-642-12912-4
Springer Heidelberg Dordrecht London New York

Cover design: WMXDesign GmbH, Heidelberg

Printed on acid-free paper

Springer is part of Springer Science+Business Media (www.springer.com)

Volume Editors

Prof. Axel H.E. Müller

Universität Bayreuth
Lehrstuhl für Makromolekulare Chemie II
Universitätsstr. 30
95440 Bayreuth
Germany
axel.mueller@uni-bayreuth.de

Prof. Hans-Werner Schmidt

Universität Bayreuth
Lehrstuhl für Makromolekulare Chemie I
Universitätsstr. 30
95440 Bayreuth
Germany
hans-werner.schmidt@uni-bayreuth.de

Editorial Board

Advances in Polymer Sciences
Also Available Electronically

Advances in Polymer Sciences is included in Springer's eBook package *Chemistry and Materials Science*. If a library does not opt for the whole package the book series may be bought on a subscription basis. Also, all back volumes are available electronically.

For all customers who have a standing order to the print version of *Advances in Polymer Sciences*, we offer the electronic version via SpringerLink free of charge.

If you do not have access, you can still view the table of contents of each volume and the abstract of each article by going to the SpringerLink homepage, clicking on "Browse by Online Libraries", then "Chemical Sciences", and finally choose *Advances in Polymer Science*.

You will find information about the

- Editorial Board
- Aims and Scope
- Instructions for Authors
- Sample Contribution

at springer.com using the search function by typing in *Advances in Polymer Sciences*.

Color figures are published in full color in the electronic version on SpringerLink.

Aims and Scope

The series *Advances in Polymer Science* presents critical reviews of the present and future trends in polymer and biopolymer science including chemistry, physical chemistry, physics and material science. It is addressed to all scientists at universities and in industry who wish to keep abreast of advances in the topics covered.

Review articles for the topical volumes are invited by the volume editors. As a rule, single contributions are also specially commissioned. The editors and publishers will, however, always be pleased to receive suggestions and supplementary information. Papers are accepted for *Advances in Polymer Science* in English.

In references *Advances in Polymer Sciences* is abbreviated as *Adv Polym Sci* and is cited as a journal.

Special volumes are edited by well known guest editors who invite reputed authors for the review articles in their volumes.

Impact Factor in 2008: 6.802; Section "Polymer Science": Rank 2 of 73

Preface

A promising approach and ongoing challenge in macromolecular science exploits the formation of complex ordered structures by (macro)molecular self-assembly. The various examples of biomacromolecules in nature impressively demonstrate how a well-defined primary molecular structure can lead to a rich complexity of structure and function on a hierarchy of different length scales. Our growing ability to control molecular structures in synthetic macromolecules opens new pathways toward novel complex macromolecular architectures, where hierarchical organization and long-range interactions enable to design new functions and tailor physical and chemical properties. The combination and interplay of these well-balanced "internal fields" with "external fields" such as mechanical, electrical, magnetic, electromagnetic, and/or the interactions with surfaces are a powerful tool to create defect-free and perfectly ordered macromolecular structures on a macroscopic scale, leading to new material properties, processes, and applications.

Therefore, the scientific understanding, control and manipulation of macromolecules, and the implementation in devices and applications of higher complexity are of critical importance for advances and breakthroughs of emerging technologies. Complex macromolecular systems play a key role in technology areas such as the generation, conversion, storage, and conservation of energy, organic electronics, photonics, information storage, and communication and display technology.

Within this context, the German Research Foundation (Deutsche Forschungsgemeinschaft) established in 1998 at the University of Bayreuth the Collaborative Research Centre (Sonderforschungsbereich) SFB 481 on *Complex Macromolecular and Hybrid Systems in Internal and External Fields*. The basis for this Collaborative Research Centre was the long-term interdisciplinary research focus on "Macromolecular and Colloid Research" at the University of Bayreuth. This focus included research groups from macromolecular chemistry, physical chemistry, experimental physics, theoretical physics, and materials science & engineering. After a maximal funding period of 12 years, members of SFB 481 now present highlights of their scientific achievements in the form of review articles in Volumes 227 and 228 of *Advances in Polymer Science*.

The first volume contains review articles on electric field-induced effects on block copolymer microdomains (by H. Schoberth, V. Olszowka, K. Schmidt, and A. Böker); on experiments and simulations of the nanopattern evolution in block copolymer thin films (by L. Tsarkova, G.J.A. Sevink, and G. Krausch); on non-lithographic approaches for topographical structuring of polymer surfaces utilizing external mechanical fields and buckling instabilities (by A. Schweikart, A. Horn, A. Böker, and A. Fery); on a generalization of the usual hydrodynamic description of destabilization and reorientation of layered systems under shear flow (by D. Svenšek and H.R. Brand); on the thermal diffusion in polymer blends, focusing on experimental and theoretical investigations on the patterning and structure formation processes (by W. Köhler, W. Zimmermann, and A. Krekhov); and on foaming of polymer blends, microstructured and nanostructured by triblock terpolymers (by H. Ruckdäschel, P. Gutmann, V. Altstädt, H. Schmalz, and A.H.E. Müller).

The second volume provides review articles on recent advances in polyelectrolyte stars and cylindrical brushes (by Y. Xu, F. Plamper, M. Ballauff, and A.H.E. Müller); on various aspects of interfacial self-assembly of nanoparticles in liquid–liquid interfaces and in block copolymers (by N. Popp, S. Kutuzov, and A. Böker); on the materials development and photophysics of azobenzene-containing block copolymers as potential media for reversible volume holographic data storage and azobenzene-containing molecular glasses for the controlled formation of surface relief gratings (by H. Audorff, K. Kreger, R. Walker, D. Haarer, L. Kador, and H.-W. Schmidt); on donor–acceptor block copolymers with nanoscale morphology for photovoltaic applications (by M. Sommer, S. Hüttner, and M. Thelakkat); and on recent advances in the improvement of polymer electret films with potential applications in microphones, sensor devices, electret filters, and in energy-harvesting devices (by D.P. Erhard, D. Lovera, C. von Salis-Soglio, R. Giesa, V. Altstädt, and H.-W. Schmidt).

On behalf of all members of the Collaborative Research Centre 481 on *Complex Macromolecular and Hybrid Systems in Internal and External Fields*, we wish to thank the Deutsche Forschungsgemeinschaft for financial and administrative support, the voluntary reviewers of the proposals for their invaluable judgment and advice, the Bavarian State Ministry of Sciences, Research and the Arts, and the University of Bayreuth for their ongoing support to continuously develop and strengthen the interdisciplinary research focus on "Macromolecular and Colloid Research" at the University of Bayreuth. Undoubtedly, all of these measures helped to advance the impact and international visibility.

Bayreuth, April 2010

Hans-Werner Schmidt
Axel Müller

The authors would like to dedicate these two volumes of Advances in Polymer Science to their colleagues and friends, Prof. Raimund Stadler and Prof. Lorenz Kramer, members of this Collaborative Research Centre, who passed away much too early. Their scientific vision and inspiring research had a great impact on progress and success of this Collaborative Research Centre.

Contents

Adv Polym Sci (2010) 228: 1–38
DOI:10.1007/12_2009_24

Published online: 31 October 2009

Polyelectrolyte Stars and Cylindrical Brushes

Youyong Xu, Felix Plamper, Matthias Ballauff, and Axel H.E. Müller

Abstract This review summarizes recent work in the field of polyelectrolyte stars and cylindrical brushes. Synthesis of anionic and cationic polyelectrolyte stars and cylindrical brushes by different synthetic strategies is introduced. Solution properties such as titration, osmotic pressure, and inter polyelectrolyte complexes (IPECs) of the stars and brushes are briefly discussed. We also highlight the use of core–shell polyelectrolyte brushes as the template for the preparation of superparamagnetic hybrid cylinders.

Keywords Core–shell structures · Cylindrical brushes · Polyelectrolytes · Star · Superparamagnetic

Contents

A.H.E. Müller (✉), Youyong Xu, and Felix Plamper
Makromolekulare Chemie II, Bayreuther Zentrum für Kolloide und Grenzflächen, Universität Bayreuth, 95440 Bayreuth, Germany
e-mail: axel.mueller@uni-bayreuth.de

M. Ballauff (✉)
F-I2 Soft Matter and Functional Materials, Helmholtz-Zentrum Berlin für Materialien und Energie GmbH, 14109 Berlin
and
Department of Physics, Humboldt University Berlin, Germany
e-mail: matthias.ballauff@helmholtz-berlin.de

1 Introduction

The past decades have witnessed tremendous developments in nanoscience and the nanotechnologies [1, 2]. Various nanostructures [3, 4] such as nanoparticles (NPs) [5], nanowires (NWs) [6], and nanotubes [7] have been successfully prepared and widely studied. They showed interesting properties, which will be turned into real applications for the near future. In the developing of these nanostructures, soft materials like polymers have been playing an important role for the templated synthesis, surface protection, and surface functionalizations of nanostructures [8]. First, the size of the polymer chains falls in the nanometer scale, which makes them natural soft building blocks. Second, polymers show diversified topologies [9–13], which contributed greatly to the templated synthesis of nanostructures in different dimensions. Scheme 1 summarizes the general topologies of polymers. Third, polymers can also self-assemble into different superstructures, either in bulk or in solution. These superstructures broaden the possibilities of applications. Fourth, owing to their vast possibilities of functionalities, polymers can respond to different external or internal stimuli, such as pH, solvent, ions, temperature, radiation, and electric or magnetic fields [14].

Polyelectrolytes, which are macromolecules carrying covalently bound anionic or cationic groups with counterions providing electroneutrality, play vital roles in the life process, industry and daily life [15–17]. Owing to the intra- and intermolecular Coulombic interactions, they demonstrate special solution properties, which are strongly influenced by the type and concentration of salt in aqueous solution. In addition to the electrostatic forces, the flexibility of the polymer backbones and molecular architectures also show great impact on their properties [18]. Many theoretical studies on the simple linear polyelectrolytes have revealed the nature of their behavior in aqueous solutions [19, 20]. Counterion condensation in the case of rod-like polyelectrolytes has been studied intensively in recent years [21]. However, polyelectrolytes with other architectures present a more complicated problem. Only recently have polyelectrolytes with nonlinear topologies attracted considerable theoretical and experimental interests, mainly originating in their possible applications as nano-building blocks and nano-templates [18].

The synthesis of polyelectrolytes with well-defined architectures, however, has imposed many challenges to the polymer chemists. Many polymerization techniques are not tolerable to the ionic functional groups. In most cases, preparation of polyelectrolytes involves the protection and deprotection of the ionic groups in the monomer. For polyelectrolytes with different architectures, various synthetic strategies are required. Recently, we have synthesized various complex architectures containing polyelectrolytes with different nonlinear topologies, such as comb-shaped [22], hyperbranched [23–25], Janus-type [26], stars [27, 28] and brushes [29–31].

Branched polyelectrolytes, like nonionic branched polymers, show interesting properties like much lower viscosities compared to the linear analogues [17]. By using the self-condensing vinyl copolymerization (SCVCP) strategy, we have successfully synthesized cationic and anionic hyperbranched polyelectrolytes [23, 25]. The degree of branching can be tuned by the ratio of the initiator-monomer (inimer) and the monomer. The inherent disadvantage of hyperbranched polymers is the broad molecular weight distribution, which causes nonuniformity and reduces their attraction for theoretical models and critical applications, which require well-defined structures.

In theoretical models and real applications, isotropic and anisotropic architectures are demanded. Polyelectrolyte stars and cylindrical brushes are best candidates for these studies. Star polymers have a core and certain number of arms, and show spherical architectures [32], while cylindrical (molecular) brushes have long linear backbones and a large number of linear side chains, and exhibit worm-like morphologies [33]. Recent progress in controlled/living polymerization techniques has made it possible to prepare polyelectrolyte stars and cylindrical brushes with defined structures and molecular weights [34, 35]. In this review we will summarize this recent progress in the study of polyelectrolyte stars and cylindrical brushes. The focus will be put on the synthesis, solution properties, and some of their applications. The use of polyelectrolyte cylindrical brushes as templates for the building of one-dimensional hybrid nano-structures will also be highlighted.

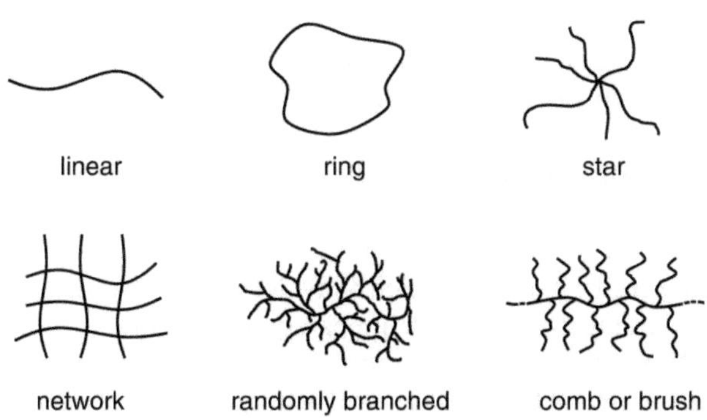

Scheme 1 General topologies of polymers

2 Polyelectrolyte Stars

Star polymers are composed of a small multifunctional core and a given number of arms [32, 36]. The core determines the arm number, and the arms determine the size of the stars. In the ideal case, the number of arms is constant throughout the sample and the arm length is the same for all the arms. However, real stars may deviate from the ideal situation. Often synthesis hardly provides both defined arm number and arm length. Hence, it is important to determine both the distribution of the arm number and arm length.

2.1 Synthesis of Polyelectrolyte Stars

Generally, there are two strategies to prepare star polymers: the core-first strategy [37–44], and the arm-first strategy [45–52]. The arm-first strategy starts with the linear arms first. Since the arms are prepared separately, many living/controlled polymerization techniques can be employed. Thus, the linear arms can be synthesized in a defined manner. Then one of the chain ends will be functionalized for further crosslinking reactions. Based on the functionalities of the chain ends, the arm-first methods can be divided into macroinitiator (MI) method and macromonomer (MM) method.

The MI method utilizes initiator-functionalized linear chains, which initiate the polymerization and crosslinking of a difunctional monomer (e.g., divinylbenzene). The active chain ends also attack the neighboring linear chains ends, and a core with crosslinked microgel is formed. In the meanwhile, certain numbers of linear chains are attached to the core. However, it is always difficult to obtain star polymers with narrow distribution of arm numbers. Quite often, many linear polymers are not attached to the core, which leads to problems in the course of the purification and for finally applications. By using multifunctional coupling agents, it is possible to get stars with uniform arm numbers. But the purification process is always unavoidable and difficult.

The MM method was recently developed by Matyjaszewski et al. [53]. The linear polymer chains are capped with monofunctional polymerizable groups. By copolymerization of the MM and coupling agents like divinylbenzene initiated from a small initiator, quite uniform star polymers with narrow molecular weight distributions can be obtained in a very high yield. Scheme 2 describes the difference between MI and MM methods.

There are only a few cases in which polyelectrolyte stars have been prepared by the arm-first strategy. Qiao et al. prepared pH responsive poly(acrylic acid) stars by the MI method using atom transfer radical polymerization (ATRP), which was used to form layer-by-layer (LBL) polyelectrolyte multilayers with linear cationic polyelectrolytes [54]. Matyjaszewski et al. obtained cationic poly[2-(dimethylamino)ethyl methacrylate] (PDMAEMA) stars and anionic PAA stars also using the MI method, which formed all-star LBL layers [55]. Ishizu et al. obtained

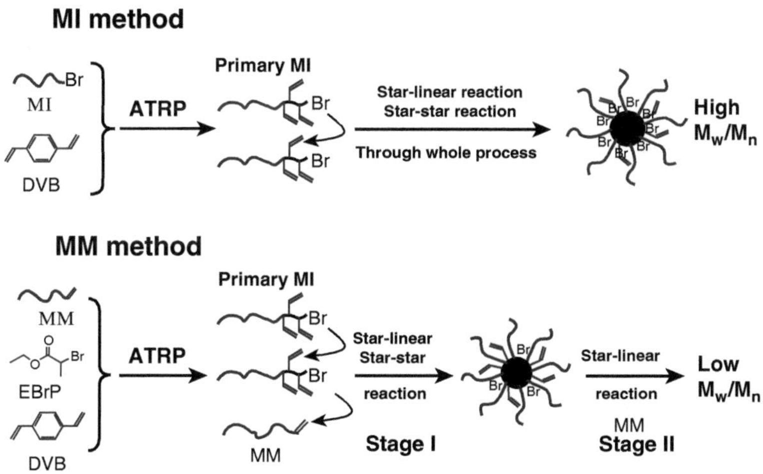

Scheme 2 MI and MM methods for the arm-first strategy [53]. Reprinted by permission of ACS

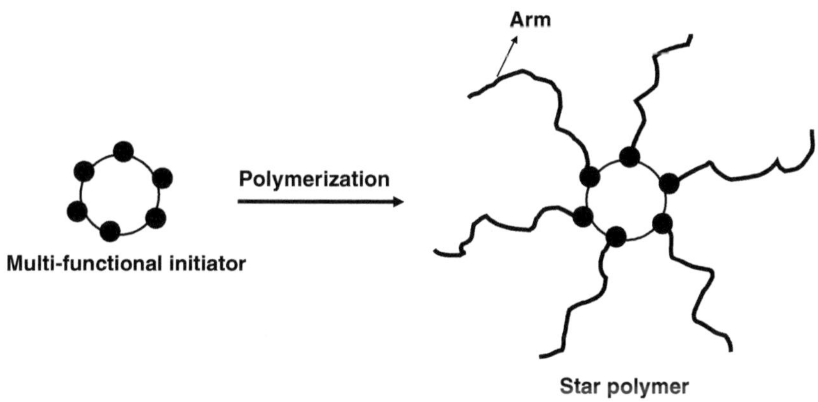

Scheme 3 Core-first strategy for the synthesis of star polymers

poly(methacrylic acid) (PMAA) stars using the same strategy [56]. Müller et al. used anionic polymerization of *tert*-butyl acrylate and star-coupling with ethyleneglycol dimethacrylate, followed by hydrolysis to obtain PAA stars [17, 57].

Based on the findings of the arm-first strategy, it is possible to obtain stars with well-defined arm length. The main problem of this strategy is the arm number distribution. Moreover, purification may cause many difficulties in the synthesis. In contrast, the core-first strategy requires multifunctional initiators and further polymerization initiated from the core. This is shown in Scheme 3. The maximum arm numbers of the stars are determined by the number of functionalities in the core. In the ideal case, the initiating efficiency of the core is close to unity, which will produce well-defined stars with precise numbers of arms. However, due to the steric hindrance and the limit of the polymerization techniques, it can be difficult to obtain full initiating efficiency.

The first step for the core-first stars is the synthesis of multifunctional initiators. Since it is difficult to prepare initiators that tolerate the conditions of ionic polymerization, mostly the initiators are designed for controlled radical polymerization. Calixarenes [39, 58–61], sugars (glucose, saccharose, or cyclodextrins) [62–68], and silsesquioxane NPs [28, 69] have been employed as cores for various star polymers. For the growth of the arms, mostly controlled radical polymerizations were used. There are only very rare cases of stars made from nitroxide-mediated radical polymerization (NMRP) [70] or reversible addition-fragmentation chain transfer (RAFT) techniques [71, 72]. In the RAFT technique one has to differentiate between approaches where the chain transfer agent is attached by its R- or Z-function. ATRP is the most frequently used technique to build various star polymers [27, 28].

An important issue in controlled radical polymerizations is radical–radical coupling. Due to the high local radical concentrations around the initiators, this may cause intra- or inter-molecular coupling of chain ends. When it happens intra-molecularly, loops of arms are formed. When it happens inter-molecularly, star–star coupling or even networks are generated. In order to obtain star polymers with defined architectures, measures should be taken to avoid these radical couplings in the polymerization process (see below).

When it comes to the polyelectrolyte stars prepared by the core-first strategy, almost all of them have been prepared by ATRP [44, 68, 73–78]. We have been active in the preparation and characterization of cationic and anionic star polymers using the core-first strategy by ATRP [27, 28, 79–82]. Anionic PAA stars were prepared by multiple steps [27]. First, multifunctional ATRP initiators with 5, 8, and 21 initiating sites were prepared by the esterification of glucose, saccharose, and β-cyclodextrin, respectively, with the α-bromoisobutyrate function. Second, the growth of poly(*tert*-butyl acrylate) (PtBA) by ATRP took place from the initiators. By using different molar ratios of monomer to initiation sites and varying the monomer conversion, PtBA stars with different arm lengths could be obtained. The PtBA stars were analyzed in terms of their molecular weight distribution by gel permeation chromatography (GPC) with viscosity detection. The number-average molecular weights, M_n, obtained are in agreement with the values calculated from conversion and monomer to initiator ratio. Finally, the PAA stars were obtained by removing the *tert*-butyl group of the PtBA stars via treatment with trifluoroacetic acid. The rather mild conditions provide almost quantitative elimination of isobutylene, yielding $(PAA_n)_x$ (n equals number-average degree of polymerization per arm, whereas x assigns the number average arm number). The ester groups used to attach the arms to the core stay almost completely intact, as shown by aqueous GPC. It is possible to methylate the PAA stars for the MALDI-TOF mass spectrometry analysis and the molecular weight values correlate with the theoretical ones. The synthetic procedures for the PAA stars and typical initiator are presented in Scheme 4.

In order to know the true arm numbers and arm lengths of the PAA stars, we cleaved the PAA arms from the core by alkaline treatment for further analysis of the arm length and arm length distribution. It is found that the lengths of the cleaved arms do coincide with the expected values. This has been verified by

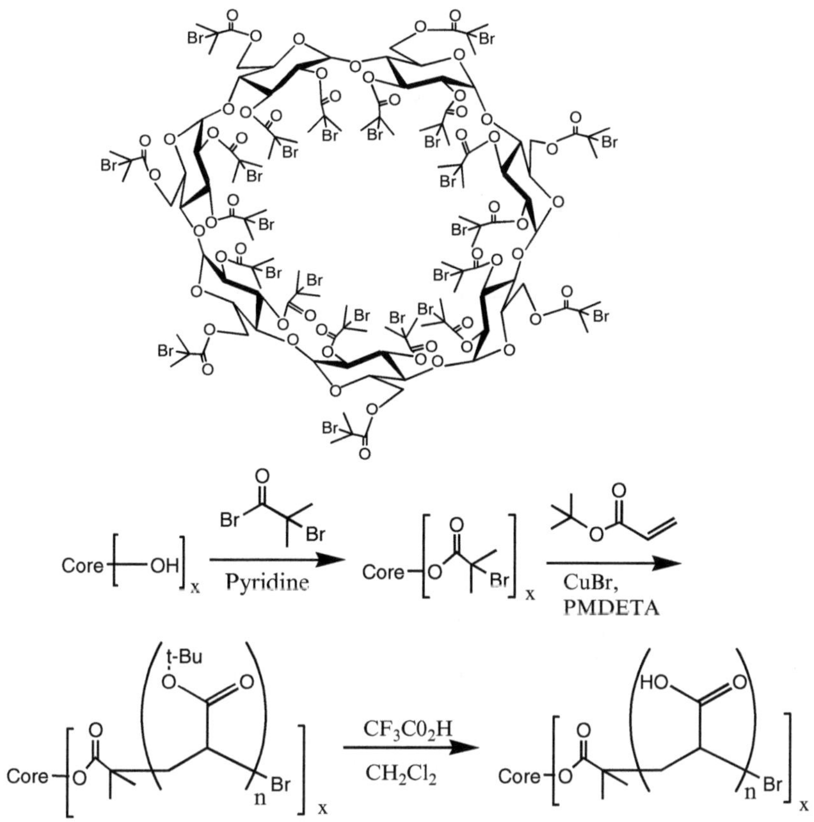

Scheme 4 Typical multifunctional initiators based on β-cyclodextrin and the synthetic procedures for PAA stars

both endgroup determination of the cleaved PAA from NMR measurements and MALDI-TOF measurements of the methylated PAA arms. This indicates that the initiation site efficiency is close to unity and the resulting PAA stars carry the arms of designed values.

In a similar manner, PDMAEMA stars were also prepared [28]. The same initiators were used for the synthesis. In order to achieve a higher number of arms, another new type of initiator was employed [83], namely silsesquioxane NPs bearing a high number of hydroxyl functions. On average, 58 initiation sites could be introduced per molecule. Using these particles, moderately polydisperse star polymers ($PDI{\sim}1.4$) with a certain arm number distribution were obtained.

The PDMAEMA stars are weak cationic polyelectrolytes. But they can be easily transformed into strong cationic polyelectrolytes poly{[2-(methacryloyloxy)ethyl] trimethylammonium iodide} (PMETAI) stars by quaternization with methyl iodide. Asymmetric field flow fractionation (AFFF) measurements showed that the silsesquioxane core remained intact. The PMETAI brushes were also subjected to

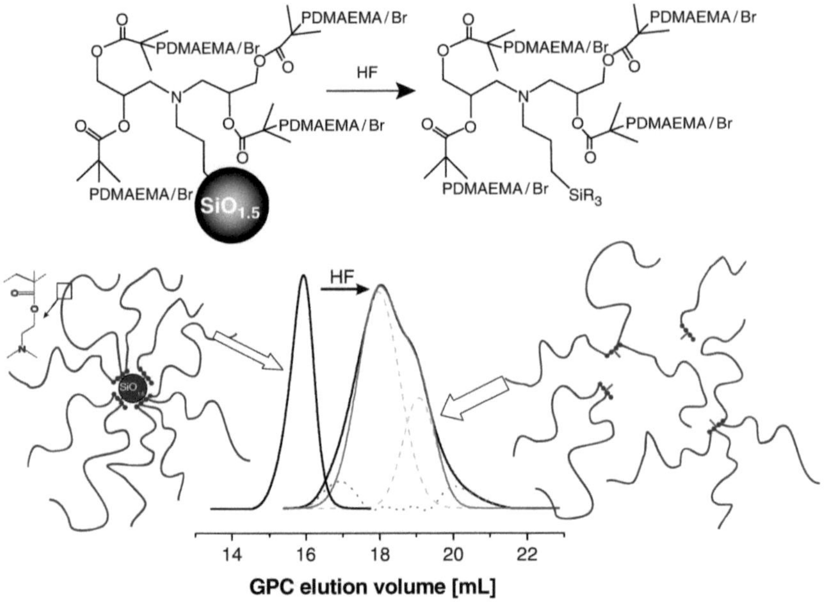

Scheme 5 Cleavage process of the silsesquioxane based PDMAEMA stars and the GPC analysis of the stars and cleaved segments [28]. Reprinted by permission of ACS

alkaline cleavage reactions. The initiation site efficiency was considerably lowered (only 0.3–0.7), especially for the initiators with a high number of functionalities. In the case of the stars with silsesquioxane cores, the initiation site efficiency could be extracted by both conventional cleavage of the arms and by a statistical method: star-like fragments were detached from the core by HF and analyzed by GPC. The results were in good agreement with those from the conventional cleavage reactions. Scheme 5 depicts the cleavage and GPC analysis.

It is possible to observe the morphologies of the cationic PMETAI stars in solution by cryogenic transmission electron microscopy (cryo-TEM). Figure 1 clearly shows the distinct and well-separated structures in the vitrified solution. In some cases it is even possible to see the silsesquioxane core [28].

2.2 *Titration Curves*

The degree of ionization of weak polyelectrolytes depends on the degree of neutralization, α, and therefore on the pH. To monitor the ionization behavior of the weak polyelectrolytes, one can measure the pH while adding strong acid or base solution. The degree of neutralization α can be calculated from the added amount of acid or base. An important value is the pK_a or pK_b, which denotes the negative decadic logarithm of equilibrium constant of dissociation of acids or protonation of bases.

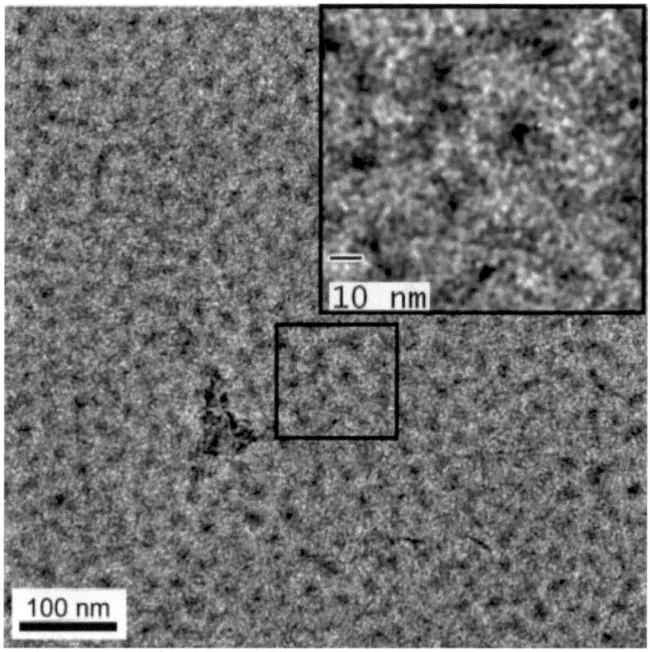

Fig. 1 Cryo-TEM image of the PMETAI stars in 2 wt% aqueous solutions [28]. Reprinted by permission of ACS

For polyelectrolytes, with a large number of acid or base functions these values are apparent only.

For the PAA stars, they were titrated against NaOH solutions [27]. Figure 2 displays the pH dependence on the degree of neutralization, $\alpha = [\mathrm{Na}^+] / [\mathrm{COOH}]_0$, where $[\mathrm{COOH}]_0$ is the total concentration of carboxyl and carboxylate groups and $[\mathrm{Na}^+]$ is equal to the amount of added NaOH. In general, the titration curves of the PAA stars are similar to that of the linear PAA. It can be seen that with constant arm length, but increasing arm numbers, the titration curves shift to higher pH. Thus, the apparent pK_a values are higher for the stars with larger arm numbers. This is in good agreement with the theoretical predictions that relate the shift of the $p\mathrm{K}_a$ to the confinement of the counterions within the polyelectrolyte star [84, 85]. However, with constant arm number but increasing arm length, the apparent pK_a values even decrease (see the inset of Fig. 2). When increasing the arm length, the mean segment density within the PAA stars decrease and less counterions are confined within the star.

Similar findings were observed for the titration of PDMAEMA stars by HCl solutions [80]. Due to the protonated PDMAEMA units and the confinement of counterions inside the PDMAEMA stars, further added HCl decreased the pH, but the protonation of the inner part of the stars needs a lower pH. Thus, the apparent pK_b values are higher than that of the linear PDMAEMA, and increase with the number of arms.

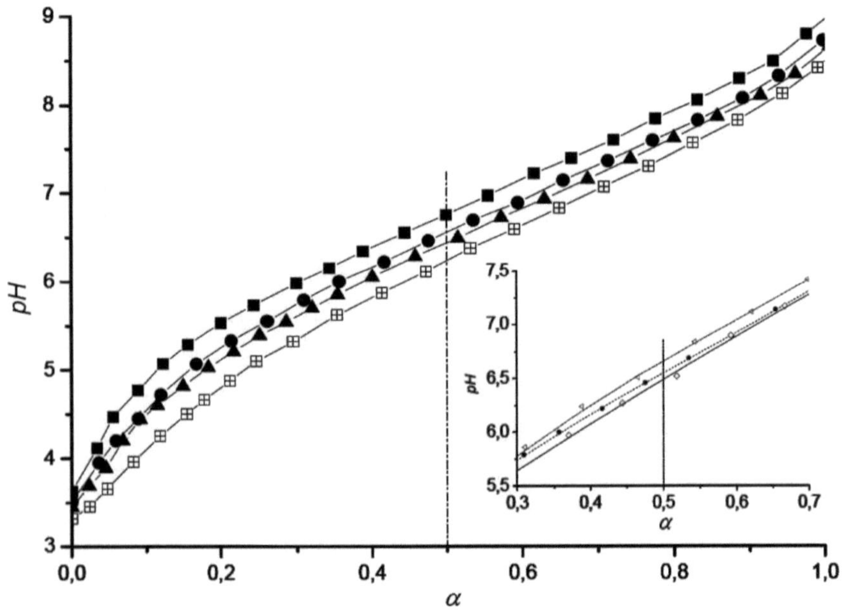

Fig. 2 Potential titration curves of the PAA stars $(PAA_{100})_{21}$ (*filled squares*), $(PAA_{100})_8$ (*filled circle*), $(PAA_{100})_5$ (*filled triangles*), and linear PAA_{100} (*open squares*). *Inset* is the cutout of $(PAA_{75})_8$ (*open triangles*), $(PAA_{160})_8$ (*open diamonds*) and $(PAA_{100})_8$ (*filled circles*) [86]. Reprinted by permission of Wiley-VCH

2.3 *Osmotic Pressure*

The osmotic pressure is one of the colligative properties, which depends on the molar concentration of the particles present in the solution. For the polyelectrolytes, however, only the counterions contribute to the osmotic pressure because their number is much larger than the number of macroions. For the polyelectrolyte, the real osmotic pressure is much lower than that calculated by the van't Hoff law because a large portion of the counterions is confined around the polyelectrolyte chains [84, 85, 87–89]. The ratio between the real osmotic pressure and the theoretical calculated one, the osmotic coefficient, φ, is a direct measure of the fraction of non-confined counterions.

For weak polyelectrolytes, the osmotic coefficient depends on the degree of neutralization and the concentration of the polyelectrolytes. We have investigated the osmotic coefficient of the PAA stars [27]. Only a low degree of neutralization was obtained under these conditions. Figure 3 shows that the osmotic coefficient is less than 0.4, indicating strong confinement of the counterions in the PAA stars. For a constant arm length, the osmotic coefficient decreases with increasing arm number. Again, this can be attributed to the stronger confinement due to the higher segment density in the PAA stars with higher arm numbers. Up to now, however, the experimental data cannot be explained by theory in a quantitative manner [90].

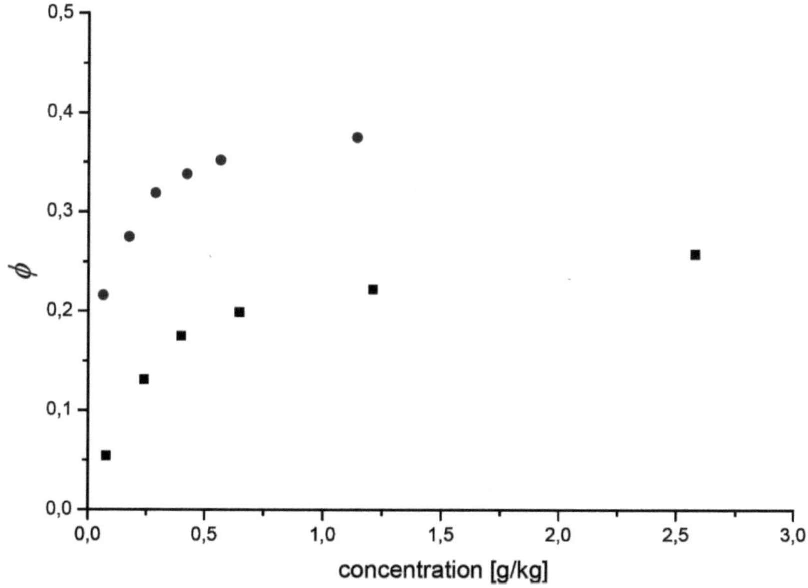

Fig. 3 Osmotic coefficient of $(PAA_{100})_{21}$ (*filled squares*, $\alpha = 0.24$) and $(PAA_{100})_8$ (*filled circles*, $\alpha = 0.25$) in dependence of concentration [86]. Reprinted by permission of Wiley-VCH

For comparison, the strong cationic PMETAI stars were also subject to osmotic pressure measurements [28]. Comparing the fully charged PMETAI stars with different arm numbers and arm lengths, similarly we found that with the increased charge density the osmotic coefficients decrease. But the trend is not as pronounced as it is for the PAA stars. This is again in contrast to theory as discussed above for the PAA stars. Nevertheless, the results for the osmotic coefficient show that more than 90% of the counterions are confined within the strong cationic PMETAI stars as predicted by theory (Fig. 4).

2.4 Effect of Counterions on the Dimensions of Polyelectrolyte Stars

As discussed above, the strong correlation of the counterions to the macroions leads to a remarkable osmotic pressure within the macroions, in this case the polyelectrolyte stars. To decrease this pressure, the arms of the polyelectrolyte stars are strongly stretched in dilute solution in the absence of added salt. However, if salt is added to the solution, the osmotic pressure difference will be reduced greatly, and the stretching of the arms will be diminished dramatically. Under these conditions the size and conformation of the polyelectrolyte stars will be comparable to the solution structure of uncharged star polymers.

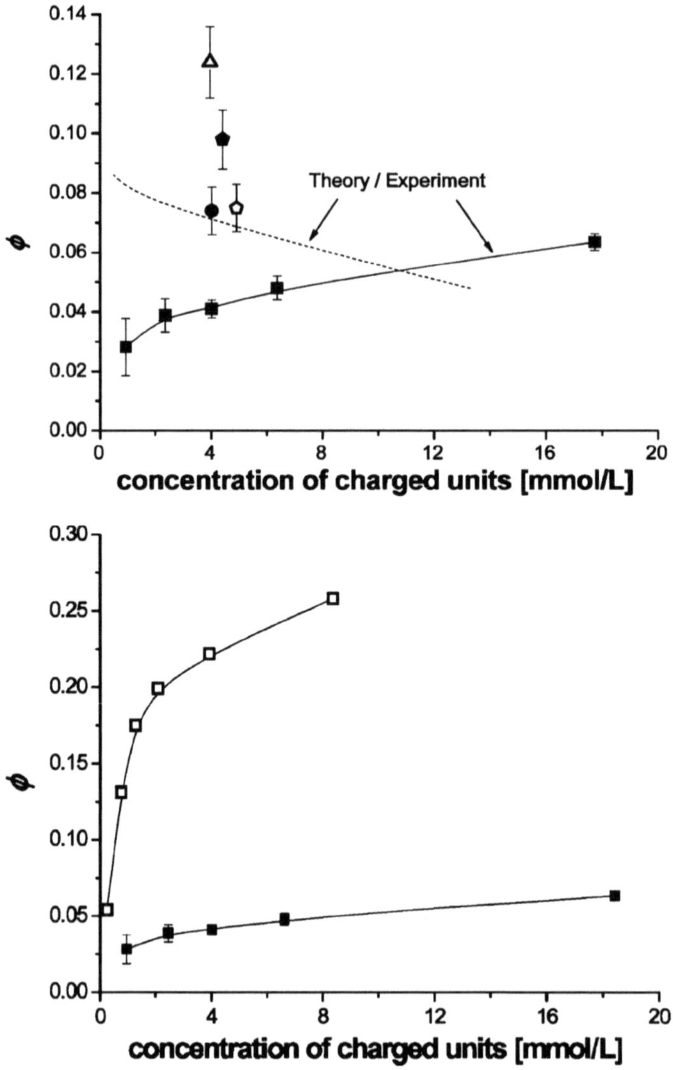

Fig. 4 Osmotic coefficients of star $(PMETAI_{170})_{18}$ (*filled squares*), $(PMETAI_{170})_{9.5}$ (*filled circles*), $(PMETAI_{110})_{5.4}$ (*open pentagons*), $(PMETAI_{170})_{5.6}$ (*filled pentagons*) and $(PMETAI_{100})_{3.1}$ (*open triangles*). *Dashed line*: theoretical dependence for the fully ionized star with 18 arms and $DP_{arm} = 170$. *Bottom*: comparison of $(PMETAI_{170})_{18}$ with PAA star $(PAA_{100})_{21}$ with $\alpha = 0.24$ (*open squares*) [28]. Reprinted by permission of ACS

The influence of added monovalent counterions to the strong cationic PMETAI stars was studied first (see Fig. 5) [28]. Dynamic light scattering (DLS) measurements of the PMETAI stars solutions with different ionic strength (NaCl and NaI solutions) show that the hydrodynamic radius, R_h, of the stars decreases with the increasing concentration of NaCl. At low ionic strength, the R_h of the stars is around 24 nm, which is about half of the contour length of the single arm, indicating the

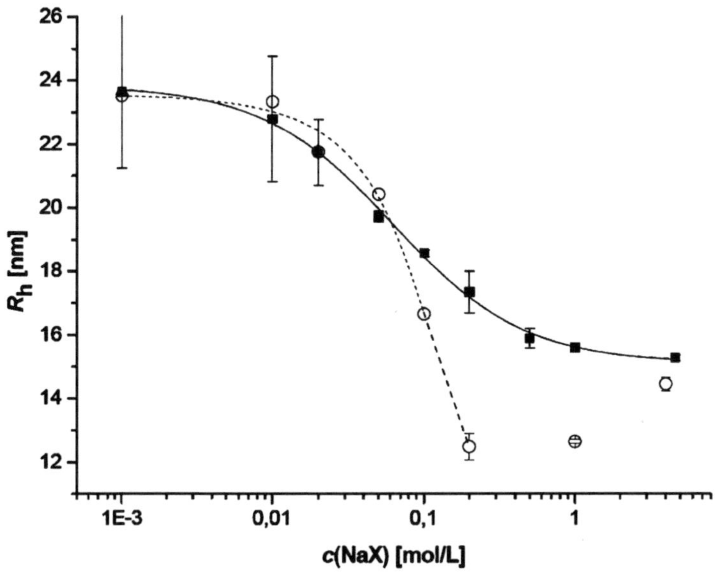

Fig. 5 Dynamic light scattering (DLS) of PMETAI brushes in NaCl (*filled squares*) and NaI (*open circles*) solutions [28]. Reprinted by permission of ACS

considerable stretching of the arms due to the high osmotic pressure within the stars. In the solution with very high ionic strength, however, the R_h is reduced to 15 nm, which is quite close to that of the unquaternized PDMAEMA stars in organic solvent. Interestingly, for the measurements in NaI solutions the stars showed salting-out and salting-in effect. When the concentration of NaI is higher than 1 M, R_h increases again, probably due to some ion-specific interactions, which was also observed for spherical cationic brushes [91].

If multivalent counterions are added to the polyelectrolyte star solutions, fast exchange of the counterions will happen [92], and a large number of monovalent counterions will be replaced by a concomitant number of multivalent ions. This will cause a remarkable decrease of the osmotic pressure, and the polyelectrolyte stars/brushes will collapse. Indeed, this has been proven by adding divalent and trivalent counterions to the PMETAI star solutions. The DLS measurements described in Fig. 6 at constant ionic strength show that both the divalent and trivalent counterions can induce the collapse of the PMETAI stars at very low concentration. When the concentration is high enough, precipitation of the polymers were observed, probably due to the linking of the stars by the multivalent counterions. It is also shown that the trivalent counterions caused more pronounced drop of the R_h than the divalent ones did, since they have more effect on the reduction of the osmotic pressure within the star [79].

The collapse of the PMETAI stars by the multivalent counterions can be used to tune the size and morphology of the stars by counterions which change their valency by an external stimulus. Hence, trivalent counterion $[Co(CN)_6]^{3-}$ can undergo

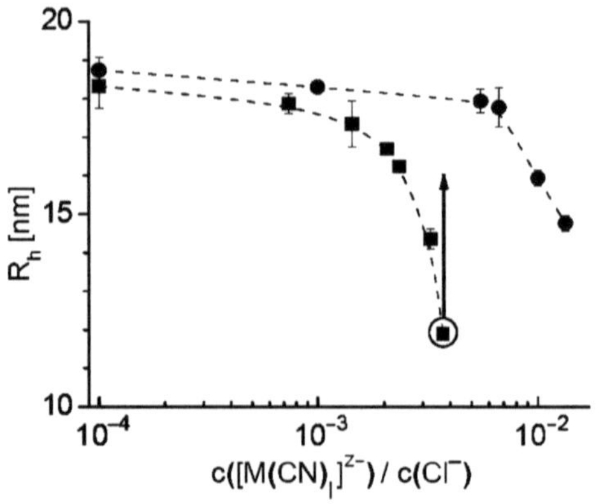

Fig. 6 DLS of PMETAI brushes with di- (*filled circles*) and trivalent (*filled squares*) counterions. The *arrow* indicates the onset of aggregation [79]. Reprinted by permission of ACS

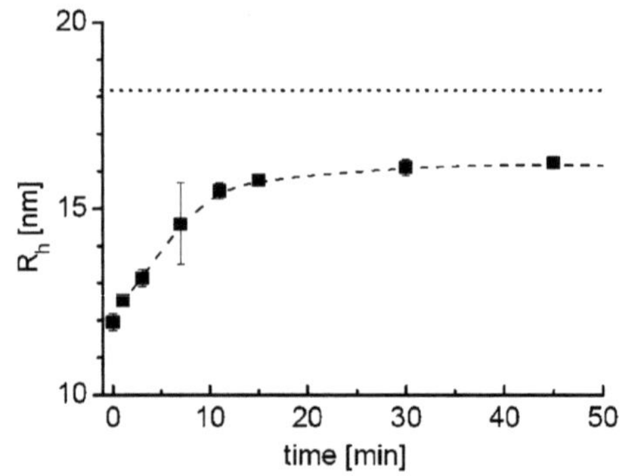

Fig. 7 UV radiation induced stretching of the PMETAI arms measured by DLS [79]. Reprinted by permission of ACS

photoaquation reaction and can be turned into one divalent counterion and one monovalent counterion. By using this effect, "nanoblossoms" were built using the PMETAI stars and $[Co(CN)_6]^{3-}$ counterions. At constant ionic strength, adding these counterions will cause the collapse of the stars as described above. When they were subjected to UV radiation, the trivalent counterions were transformed into divalent and monovalent counterions which increased the osmotic pressure within the stars. The effect of the UV radiation is shown in Fig. 7, and apparent increase of R_h was observed. Thus the arms of the stars opened, which resembles blooming of real flowers in sunlight [79] (see Scheme 6).

2.5 *Temperature Responsiveness of the PDMAEMA Stars*

Linear PDMAEMA exhibits a lower critical solution temperature (LCST) in water, that is the polymer becomes insoluble at elevated temperatures. We have

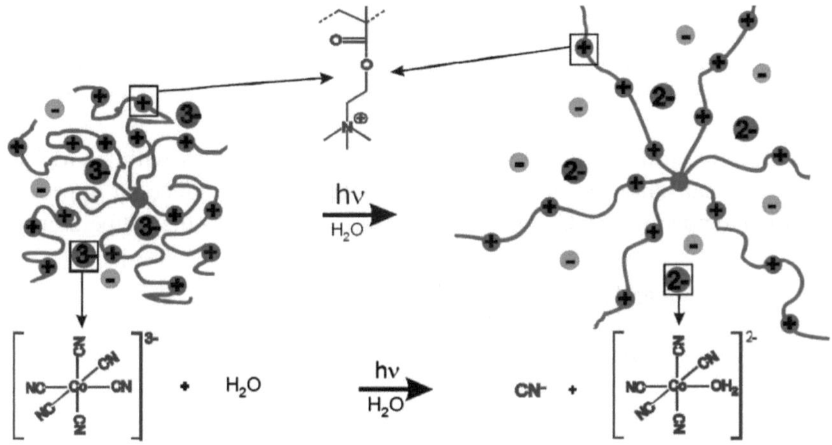

Scheme 6 Schematic mechanism of the nanoblossom [79]. Reprinted by permission of ACS

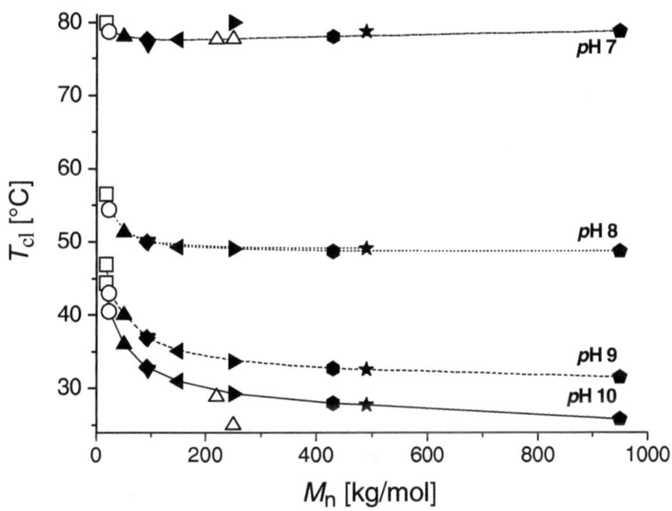

Fig. 8 Cloud points at $0.1\,gL^{-1}$ of *star-shaped* PDMAEMA in dependence of M_n and pH. The *open symbols* represent linear polymers. The *lines* are a guide to the eye [80]. Reprinted by permission of ACS

systematically investigated the thermo-responsive behavior of linear and star-shaped weak polyelectrolyte PDMAEMA. Since the pH changes with temperature, it is important to keep the pH constant for all the samples by buffer solutions [80] (Fig. 8).

We found that the observed cloud points (determined by turbidity measurements at 500 nm) only depend on the molecular weights (according to a Flory–Huggins behavior) at high pH, and that the architecture of the polymer does not show a significant impact. However, at lower pH a slight modulation is observed: in addition to the shift of the whole cloud point curve to elevated temperatures, the cloud points of stars with higher segment densities are shifted to slightly higher temperatures compared to stars with lower segment density. Figure 9 clearly shows the pH dependence of the cloud point for a linear polymer and two stars with similar DP of the arms.

Interestingly, in the buffered solutions of PDMAEMA stars with 0.1 M NaCl, addition of small amounts of multivalent counterions can induce an upper critical solution temperature (UCST), i.e., demixing upon lowering the temperature (see

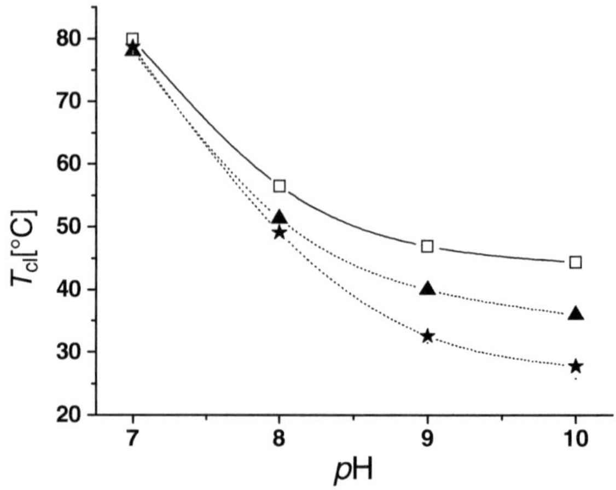

Fig. 9 Cloud points at $0.1\,\mathrm{g\,L^{-1}}$ of *linear* and *star-shaped* PDMAEMA in dependence of pH. (*Open squares*) $PDMAEMA_{108}$; (*filled triangles*) $(PDMAEMA_{100})_{3.1}$; (*filled asterisks*) $(PDMAEMA_{170})_{18}$ [80]. Reprinted by permission of ACS

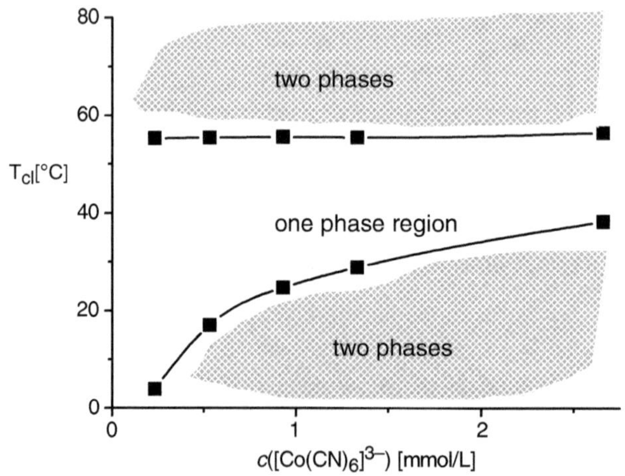

Fig. 10 Dependence of the cloud points of aqueous $(PDMAEMA_{170})_{18}$ solutions ($0.1\,\mathrm{g\,L^{-1}}$ in buffer of pH 8 + 0.1 MNaCl) on the $[Co(CN)_6]^{3-}$ concentration (*upper curve* assigns LCST-type cloud points, *bottom curve* refers to cloud points of the UCST-behavior) [81]. Reprinted by permission of ACS

Fig. 10). The UCST-type cloud points shift to higher temperatures with the increased concentration of added trivalent counterions, whereas the LCST-type cloud points hardly change. Thus, the LCST-type transition can be adjusted by pH, while the UCST-type cloud points can be adjusted by the concentration of trivalent counterions. It is also possible to switch the UCST behavior of the PDMAEMA by using the light-sensitivity of hexacyanocobaltate(III) as the trivalent counterions. Under UV illumination, the trivalent counterions are turned into a mixture of divalent and monovalent counterions and the UCST-behavior disappears [81].

2.6 Ordered Structure in Solution

It has been predicted by theory that the polyelectrolyte stars can form different kinds of ordered structures in solutions at high concentrations. Likos et al. [93, 94] pointed out the different crystal structures for polyelectrolyte stars with certain arm numbers and densities (see Fig. 11). This has been used to explain the anomalous structure with the concentrated polyelectrolyte star solutions. Ishizu et al. [95] successfully demonstrated this ordering by small angle X-ray scattering (SAXS) for concentrated PAA star solutions with arm number 30 and 97 (displayed in Fig. 12). No such phenomenon was seen for the polyelectrolyte star polymers under scrutiny here. The reason for this may be located in the finite polydispersity of the present systems that may suppress the formation of ordered phases.

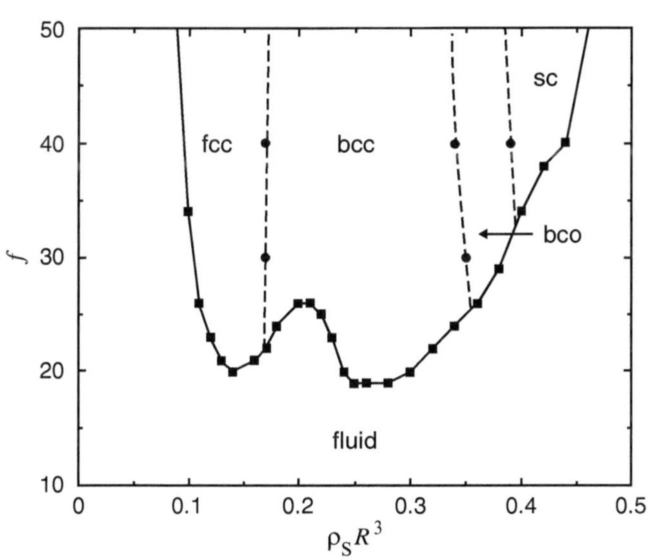

Fig. 11 Theoretical phase diagram of polyelectrolyte stars with charge ratio $\alpha = 1/3$, drawn in the density functionality plane [93]. Reprinted by permission of IOP

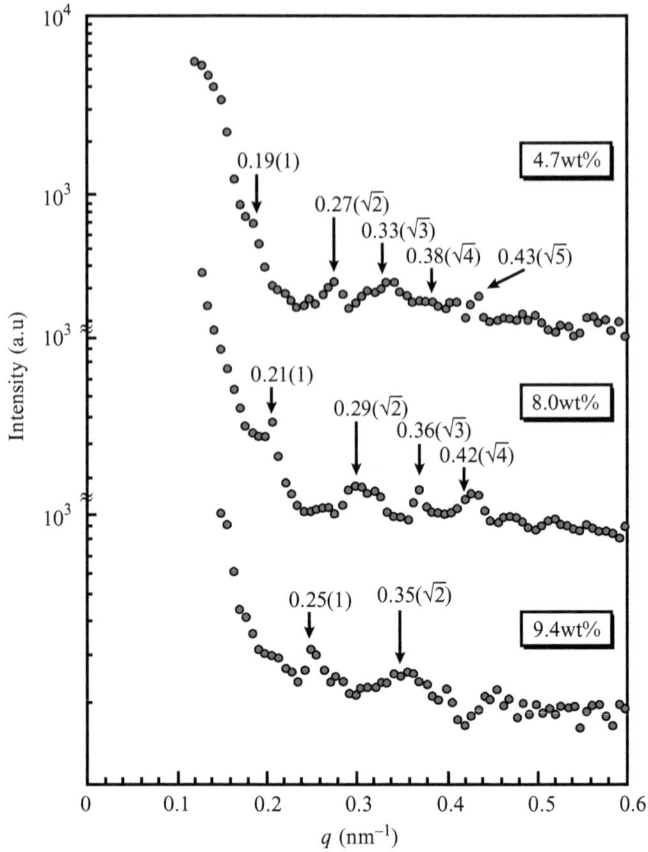

Fig. 12 SAXS intensity profiles for PAA stars with arm number = 97 and $DP_{arm} = 25$ at different concentrations [95]. Reprinted by permission of ACS

2.7 Inter-Polyelectrolyte Complexes of Polyelectrolyte Stars

Polyelectrolytes carrying different charges tend to form inter-polyelectrolyte complexes (IPECs) [96–101]. Polyelectrolyte stars, in particular, can form IPECs with oppositely charged polyelectrolytes of different architectures. We have investigated the IPECs formed by PAA stars with different arm number (5, 8, and 21) and linear cationic poly(*N*-ethyl-4-vinylpyridinium bromide) (P4VP·EtBr) [82]. It is possible to obtain water-soluble complexes. However, two populations of complex species with a large difference in size were found. The mixtures of the oppositely charged polymeric components remain transparent until their base-molar ratio exceeds a certain threshold value, which depends on the number of arms of the poly(acrylic acid) stars and the ionic strength. The small complex species are the major fraction, and are assumed to be the particles of water-soluble IPEC. The large complex species are the minor fraction, and are considered to be complex aggregates. The fractions of these complex species but not their sizes distinctly depend on the number of arms of the poly(acrylic acid) stars and the charge ratios as well as the ionic strength (See Scheme 7).

IPECs may have an application for gene delivery. Müllen et al. prepared cationic polyelectrolyte stars from a dendritic polyphenylene core, which formed stable IPECs with series of negatively charged DNA [102]. When treated with high concentration of NaCl, the DNA can be released from the complexes. Wang et al. found that oppositely charged polyelectrolyte stars can form IPECs of various morphologies such as fiber bundles and vesicles by tuning the molecular weights, charge ratios, and polymer concentrations [103]. Applying the LBL technique, Qiao et al. [54] and Paula et al. [55] have prepared pH responsive multilayers by star/linear and star/star complexes, respectively.

2.8 *Core–Shell Polyelectrolyte Stars*

For the core-first strategy it is possible to build the star block copolymers by consecutive polymerization of different monomers. There are many examples of this type of core–shell polyelectrolyte stars, most of which are amphiphilic. Tenhu et al. have prepared block copolymer stars with PMMA as the core and PAA as the shell [74, 76]. Since the core is hydrophobic and the shell is water soluble, they can self-assemble into interesting cylindrical structures in aqueous solutions. Detailed studies showed that the self-assembly behavior is greatly influenced by the arm numbers and the ionic strength [77]. The core–shell polyelectrolyte stars can also serve as the template for the synthesis of inorganic NPs. Shi et al. [104] prepared stars with PS as the core and PVP as the shell. In toluene, the stars form inverse molecular micelles. Gold NPs can be generated in the PVP part, forming the inorganic/organic hybrid materials.

2.9 *Some Applications of Polyelectrolyte Stars*

The chain ends of the star polymers can be functionalized. So it is possible to link these sites and form networks or gels of stars. Depending on the chemical nature of the chain end, the crosslinking can be permanent or stimuli-switchable. Patrickios et al. [105] have prepared crosslinked double hydrophilic star copolymers made from methacrylic acid (MAA) and poly(ethyleneglycol methacrylate) (PEGMA). Meyer et al. [106] synthesized photocrosslinkable PMAA stars, which could serve as well-defined model polyelectrolyte networks. The polyelectrolyte stars can also be used as template for some nanomaterials. Want et al. [107] prepared strong anionic polyelectrolyte stars by sulfonation of the PS stars. They were used as the template for the preparation of conducting polyaniline.

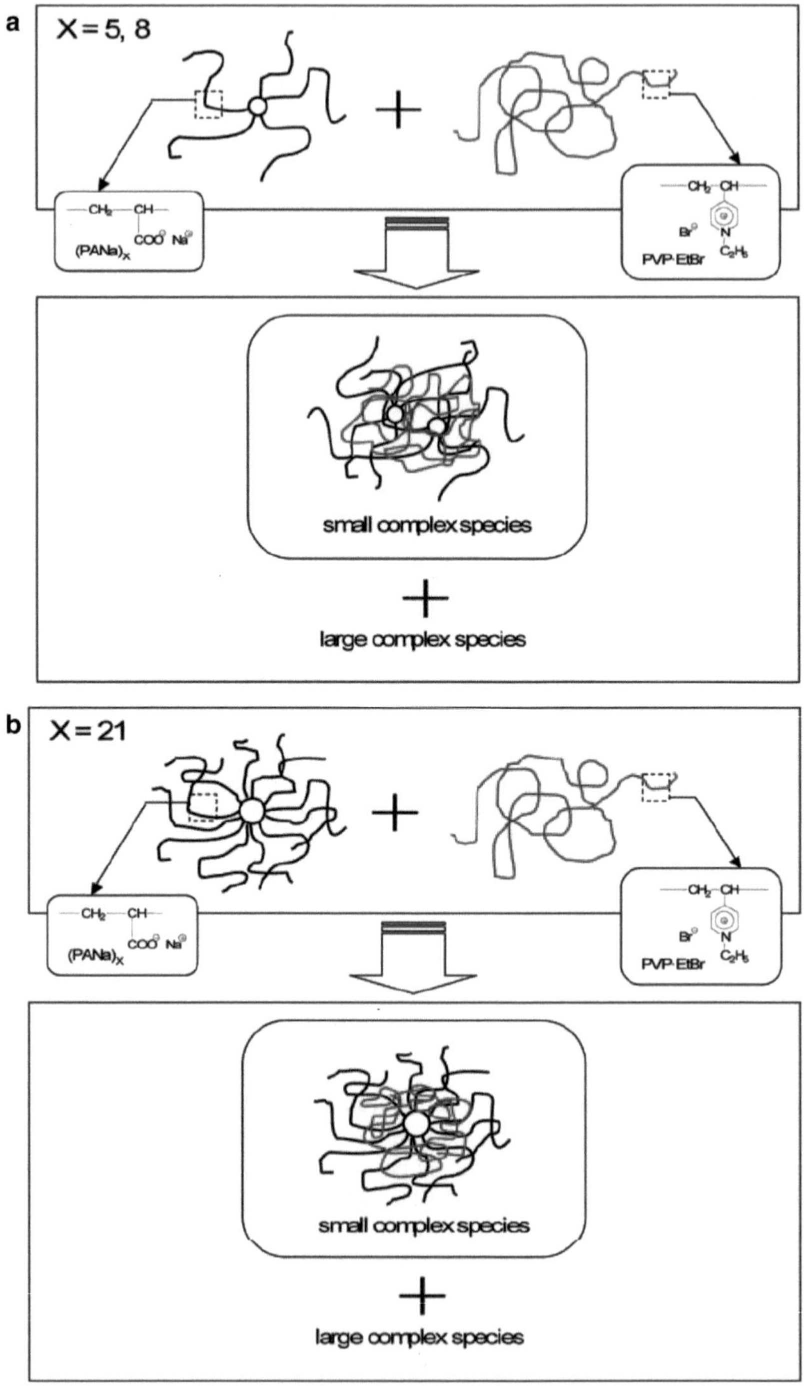

Scheme 7 Schematic description of the IPECs formed by PAA stars and P4VPB. (**a**) P4VPB and PAA stars with arm number = 5 and 8, and (**b**) P4VPB and PAA stars with arm number = 21 [82]. Reprinted by permission of ACS

3 Cylindrical Polyelectrolyte Brushes

When a linear polymer is grafted with a large number of much shorter side chains, cylindrical polymer brushes are formed [33, 108–111]. They are also denoted as "bottlebrushes" or "molecular brushes." Although most cylindrical polymer brushes contain linear side chains, dendritic or even hyperbranched space demanding grafts can also render cylindrical shapes, which leads to the so-called dendronized [112–116] and hypergrafted [117, 118] polymers, respectively. In this review, we will focus on cylindrical brushes with linear side chains. Due to their anisotropic nature in topology, they have attracted more and more research interest in their synthesis, bulk, or solution properties, as well as applications.

3.1 Synthesis of Polyelectrolyte Cylindrical Brushes

Three main strategies have been successfully developed for the synthesis of cylindrical polymer brushes: grafting through, grafting to, and grafting from. Scheme 8 shows the processes for each preparation strategy.

The grafting through strategy means the preparation of cylindrical brushes by the polymerization of preformed MMs. This was the first method in the synthesis of cylindrical brushes. For the first time, Tsukahara et al. [108] successfully obtained cylindrical brushes by the radical polymerization of MMs. Short polymer chains with a polymerizable vinyl chain end were first made by anionic polymerization and sequential end-functionalization. Further radical polymerizations of the premade MMs yielded cylindrical brushes with uniform side chains. This method was then further employed by Schmidt et al. [109, 119–121] and Ishizu et al. [122–127] to

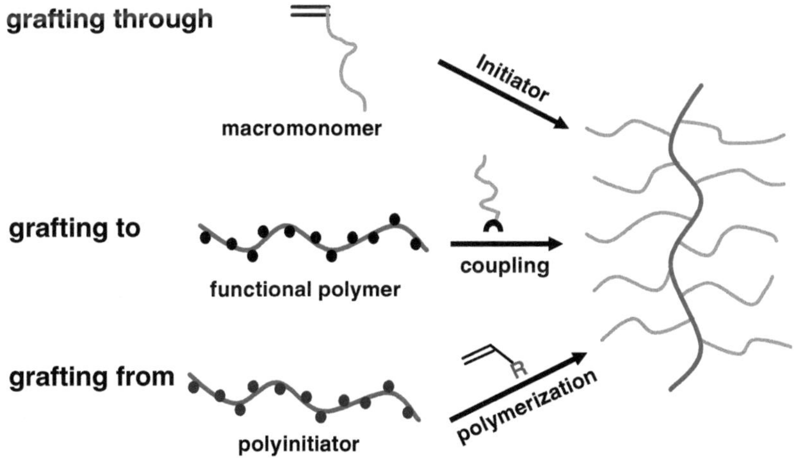

Scheme 8 Three different strategies for the preparation of cylindrical brushes

prepare cylindrical brushes with different functionalities and architectures. The most obvious merit of this method lies in the high grafting density of the side chains. However, due to the great steric hindrance caused by the MMs, it is difficult to obtain very long brushes by this method. Conventional radical polymerizations can only afford cylindrical brushes with a rather broad distribution of the backbone length. More importantly, it is very difficult to obtain complete conversion of the MMs, which results in a more difficult purification process. To remove the residual MMs from the reaction mixtures, fractionation is generally required. These, in combination, have limited the use of this strategy for the preparation of cylindrical brushes with well-defined structure.

Nevertheless, many cylindrical polyelectrolyte brushes have been prepared by this strategy. Schmidt et al. [128] prepared weak anionic PMAA brushes by the polymerization of PtBMA MM and further hydrolysis. However, fractionation was necessary to remove the remaining MMs. Wegner et al. [129] prepared PtBA MMs by ATRP and chain-end functionalization. Radical polymerization and subsequent hydrolysis reactions led to the PAA brushes. It is also possible to prepare strong anionic brushes with sulfonate groups by the sulfonation of PS brushes. However, it is difficult to avoid the side-reactions and crosslinking [128]. Using the same strategy, Schmidt et al. [119, 128] also prepared cationic poly(2-vinylpyridine) (P2VP) brushes and from there quaternized cationic brushes.

Cylindrical brushes can also be built by coupling reactions between the preformed monotelechelic polymers and preformed multifunctionalized long backbones, namely, the "grafting to" strategy [130, 131]. The advantage of this technique is that both backbone and side chains can be well defined since they are prepared separately. For example, Matyjaszewski et al. [132] prepared both backbones and grafts by ATRP and click chemistry for the coupling reactions. However, it is difficult to obtain high grafting densities with the "grafting to" method because of the high steric hindrance caused by the crowding of the side chains. Incomplete coupling reactions between the side chains and backbones cause additional problems for the purification as compared to the "grafting through" method. So far, examples of polyelectrolyte brushes prepared by this strategy are very rare.

Finally, the "grafting from" strategy was first employed to synthesize different cylindrical brushes with both well-defined backbones and side chains by Matyjaszewski et al. [133]. Generally, a narrowly distributed long backbone is first prepared via living polymerization techniques, followed by functionalization to attach initiating groups to the backbone for the further grafting polymerizations. The backbone polymer determines the length and distribution of the cylindrical brushes. To obtain very long functional backbones, different living polymerizations, such as anionic polymerization [31], ATRP [30, 133, 134], and RAFT polymerization [135] have been employed. For example, in our group [31, 136], anionic polymerization of silyl-protected 2-hydroxylethyl methacrylate (HEMA) and further esterification reactions produced long polyinitiators with DP_n up to 3,200 as shown in Scheme 9.

For the grafting of the side chains, different controlled polymerization techniques were also applied. In addition to the rare cases of ring-opening polymerizations

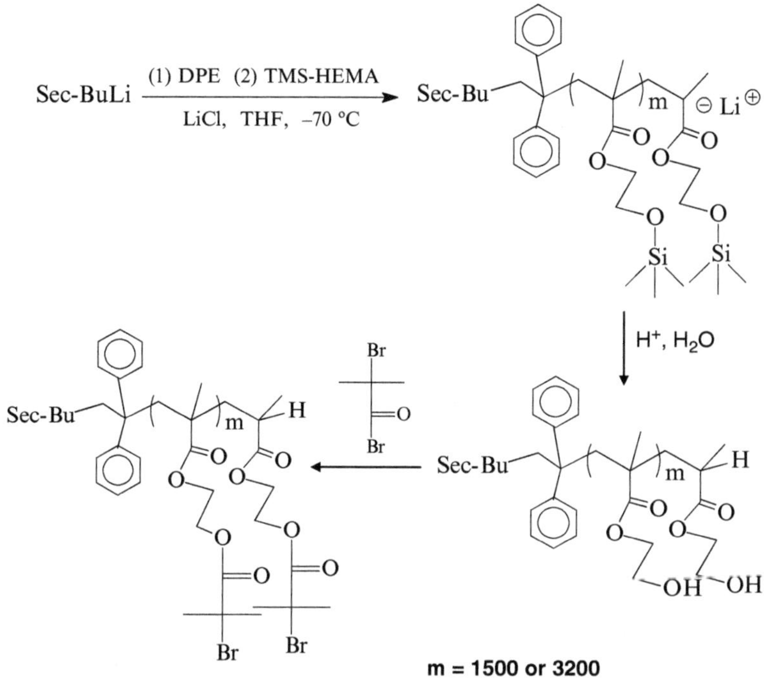

Scheme 9 Preparation of polyinitiators by anionic polymerizations and further functionalizations

(ROP) [137] and NMRP [138], ATRP has been the dominating technique for the preparation of most of the side chains [30, 31, 133, 134]. In the case of polyinitiators the situation is more complicated. Due to the very high local concentration of radicals around the linear polyinitiators, the rate of polymerization is normally higher than that for the ATRP initiated by small initiators. The concentration of the catalyst/ligand complex should therefore be kept low to avoid inter- or intramolecular coupling reactions [133]. Although few cylindrical brushes made by ATRP showed high initiating efficiency (e.g., with polystyrene side chains [30]), low initiating efficiencies were commonly reported [29, 139–141], in particular for α-methyl substituted monomers like methacrylates, due to the steric hindrance and slower initiation compared to the rate of propagation. Matyjaszewski et al. [140] reported on the improvement of the initiating efficiency by a lower monomer concentration, a higher concentration of the active Cu(I) catalyst by adding Cu(II), and halogen exchange.

So far, the "grafting from" strategy has proven to be the most successful one for the preparation of cylindrical brushes. Both anionic [30, 31, 142, 143] and cationic [29] polyelectrolyte cylindrical brushes with well-defined backbones and side chains were prepared. We have also prepared core–shell brushes with PAA as the core or shell, which will be discussed later [30, 31]. Recently we have been able to prepare strong anionic brushes with sulfonate groups by direct ATRP of ionic monomers in organic solvent [144].

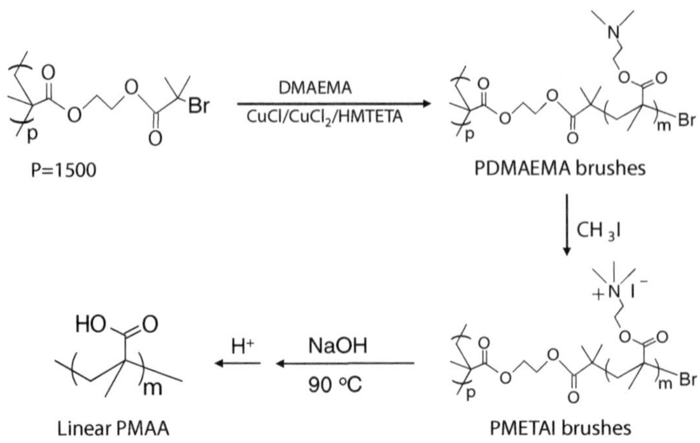

Scheme 10 Synthesis of PDMAEMA, PMETAI brushes and side chain cleavages

We also prepared weak and strong cationic cylindrical brushes using the "grafting from" strategy. PDMAEMA cylindrical brushes were prepared by the combination of anionic polymerization and ATRP using this strategy [29]. A polyinitiator with $DP_n = 1,500$ and a very low polydispersity served as the backbone and linear PDMAEMA chains were grafted from it. Several measures, such as halogen exchange, lower monomer concentrations and slower growth by adding Cu(II) ions, were taken into account to improve the grafting density. For the determination of the initiating efficiency, PDMAEMA brushes were first quaternized to their ammonium salt and were later subject to cleavage of the side chains and the ester groups in the monomer unit by strong base. This led to linear PMAA, measured by aqueous GPC. By comparing the molecular weights of the linear PMAA and that of the calculated values, the initiating efficiencies can be obtained. It is found that even when the above-mentioned measures had been taken, the initiating efficiencies were all only around 50%. Scheme 10 displays the synthesis of PDMAEMA, PMETAI brushes and the process for the cleavage of side chains.

3.2 *Morphology of the Polyelectrolyte Cylindrical Brushes*

Due to the strong repulsion of the side chains, cylindrical brushes normally exhibit worm-like morphologies in good solvent. Based on the electrostatic repulsion, polyelectrolyte cylindrical brushes, carrying a large number of charged side chains, show even more extended structures than their neutral analogues. It is possible to analyze their morphology by atomic force microscopy (AFM) on suitable substrates [29]. Figure 13a shows the worm-like morphologies of the strong cationic PMETAI brushes deposited on the negative mica surface in the dry state. Also, cryo-TEM is a powerful tool to examine the true morphologies of the brushes in solution, since

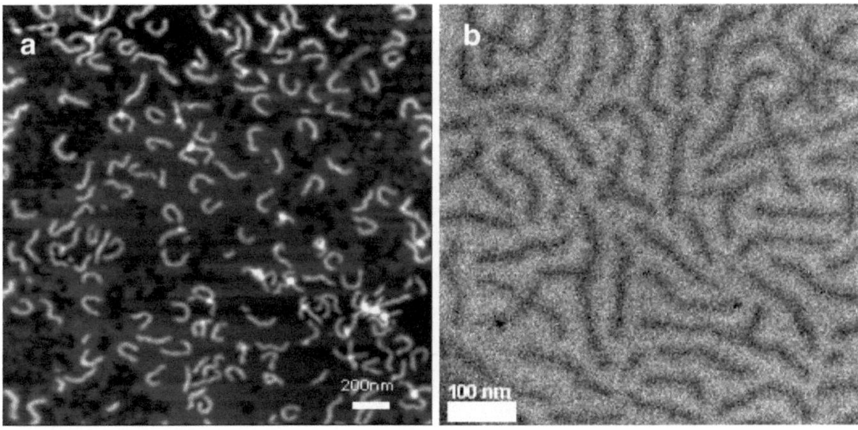

Fig. 13 (**a**) AFM height image (*Z* range 7 nm) and (**b**) cryo-TEM image of PMETAI brushes [29]. Reprinted by permission of Elsevier

there is no interaction with a surface [29, 142, 145, 146]. Figure 13b reveals the extended worm-like structures of the same brushes in aqueous solution. The size and structure determined by these two methods show similar results which in turn proves the reliability of the AFM measurements.

3.3 pH Responsiveness of Weak Cationic Brushes

PDMAEMA is a weak cationic polyelectrolyte and the apparent pK_b value for PDMAEMA stars with a large number of arms is around 8. The same behavior is expected for PDMAEMA brushes. In this case, a change of the ionization degree is expected to cause structural changes. DLS measurements of a PDMAEMA brush at different pH values were carried out [29]. All measurements were performed in the absence of salt at very low concentration $\left(0.2\,\mathrm{g\,L^{-1}}\right)$ to avoid intermolecular aggregation at high pH values. An obvious pH dependence of the apparent hydrodynamic radii is observed, as is shown in Fig. 14. The apparent hydrodynamic radii steadily decreased with increasing pH. Structural changes were directly observed by cryo-TEM measurements at different pH values. More extended structures appeared at low pH, while at high pH the brushes were extraordinarily shortened.

3.4 Effect of Added Counterions on Strong Cationic Brushes

The strong cationic PMETAI brush adopts a rather elongated conformation due to the large osmotic pressure along the brush layer, as illustrated in Fig. 12b. Addition of salt to the solution will lower the net osmotic pressure. Thus, the stretching of the

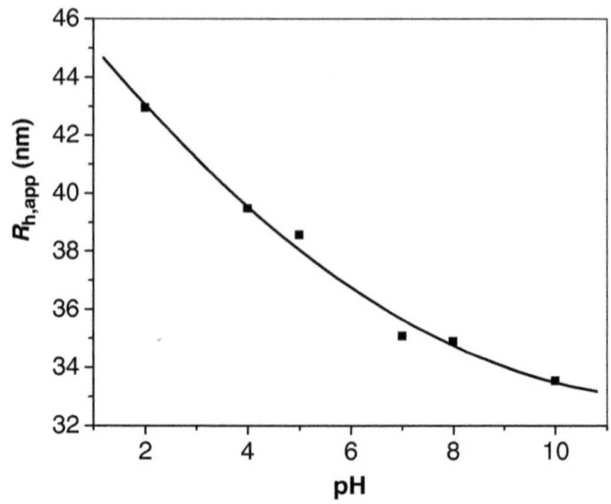

Fig. 14 Apparent hydrodynamic radii of PDMAEMA brush in $0.2\,\mathrm{g\,L^{-1}}$ aqueous solution at different pH values [29]. Reprinted by permission of Elsevier

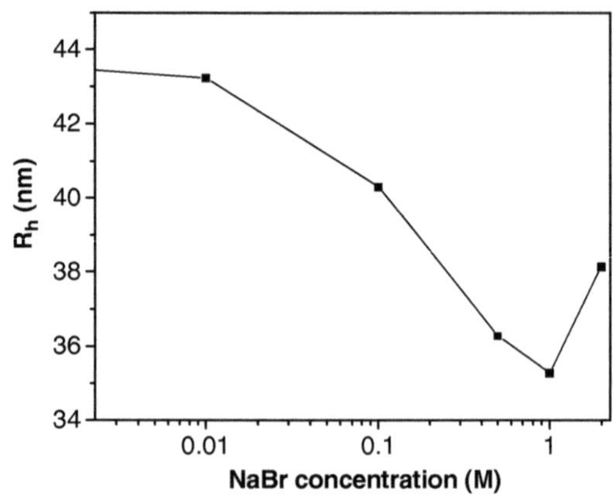

Fig. 15 Apparent hydrodynamic radii of the PMETAI brushes as a function of NaBr concentration [29]. Reprinted by permission of Elsevier

polyelectrolyte chains should be diminished and the conformation should collapse. Figure 15 shows the change of the apparent hydrodynamic radii of the PMETAI brush with increasing added NaBr concentration [29]. Increasing the NaBr concentration from 0 to 1 M leads to a remarkable decrease of R_h. This is in accordance with the findings reported on spherical polyelectrolyte brushes and polyelectrolyte stars. When the salt concentration is higher than 1 M, R_h increases again. This has also been found for the other polyelectrolyte systems. A specific ion interaction with the cationic polyelectrolytes can be invoked as the reason for this unexpected phenomenon. The collapse of the PMETAI brushes at high salt concentration was directly visualized by AFM.

When the monovalent counterions are replaced by divalent or trivalent counterions, the osmotic pressure within the polyelectrolyte architecture will decrease dramatically due to the exchange of a considerable amount of entrapped, monovalent counterions with rather few multivalent counterions. Therefore a contraction or collapse should be observed at a much lower concentration of added salt. Here we chose two cyanometalates, $[Ni(CN)_4]^{2-}$ and $[Co(CN)_6]^{3-}$, as the di- and trivalent counterions, respectively [145].

AFM measurements were performed from highly dilute solutions $(0.02\,gL^{-1})$ to avoid aggregation. The influence of the divalent counterions is studied first. The concentration of added salt was 2.2×10^{-2} mM. Given the small concentration of added salt, a molar charge ratio $Z_{-/+}$ can be defined which is given by the ratio of the added anionic charges and the cationic charges carried by the PMETAI brushes. For the case $Z_{-/+} = 0.66$, all $[Ni(CN)_4]^{2-}$ ions present in the system would not suffice to neutralize the positive charges of the cylindrical brush. Therefore the brushes still kept their worm-like structures. When the charge ratio $Z_{-/+}$ was increased to unity, most of the brushes collapsed into globular structures. However, an entirely new feature appears at high dilution when $Z_{-/+} = 0.75$. AFM measurements reveal an intermediate helical state with a pitch of ca. 25 nm. The chains twist in order to relieve a part of the strain caused by the contraction along the long axis of the molecule. In some cases this leads to a helix with several turns. One explanation of the helix formation is given on the basis of entropic arguments [147]. The helical structures can increase the overlapping volume of the brushes, and thus increase the entropy of the whole system.

The same phenomenon was observed with trivalent counterions at $Z_{-/+} = 1$. Figure 16 shows the typical AFM image of the brushes with added trivalent counterions and enlarged portions. Clear helical structures are observed. Since there is no

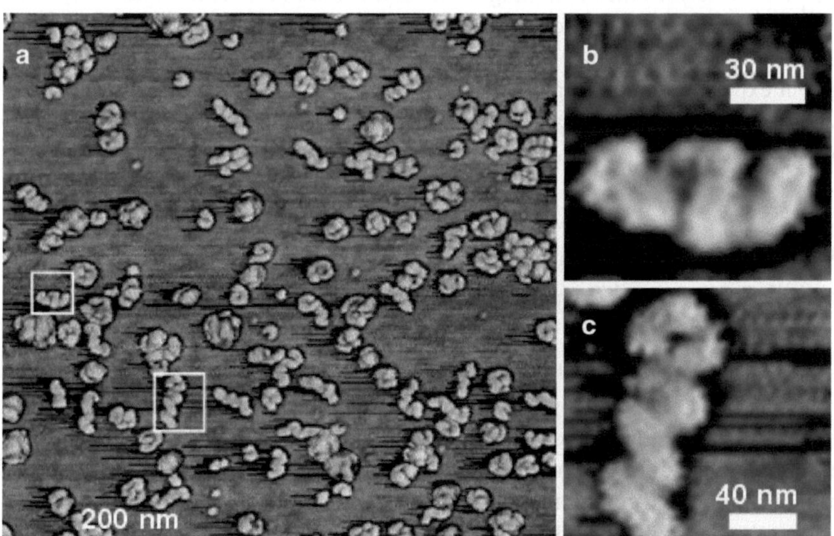

Fig. 16 (**a**) AFM height image of PMETAI brushes with added trivalent salt spin-coated on mica, $Z_{-/+} = 1$, z range 12 nm, brush concentration $0.02\,gL^{-1}$, and the trivalent salt concentration 2.2×10^{-2} mM; (**b**, **c**) show a typical right- and left-handed helix, respectively [145]. Reprinted by permission of Royal Society of Chemistry

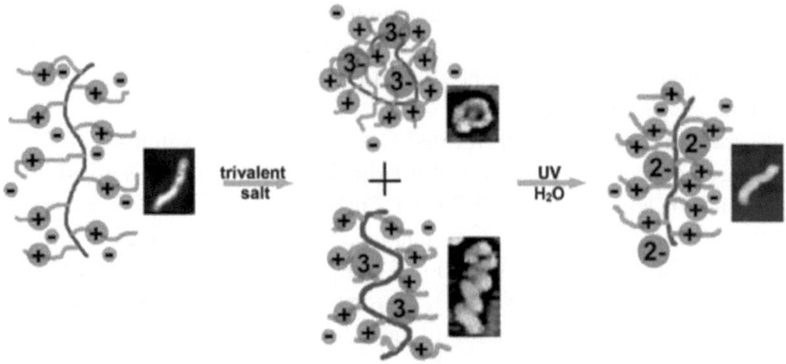

Scheme 11 Morphology change of the PMETAI brushes with added $[\mathrm{Co(CN)_6}]^{3-}$ counterions and UV radiation

chiral unit within the chains, both right-handed and left-handed helices are observed in similar numbers. Figure 16a shows that the helical conformation is an intermediate state between the stretched and fully collapsed spherical conformation.

Similar to the "nanoblossom" system discussed for star polymers [79], the photosensitive trivalent counterion $[\mathrm{Co(CN)_6}]^{3-}$ at charge ratio $Z_{-/+} = 1$, leading to a full collapse of the brushes, were illuminated with UV light, transforming them into divalent and monovalent counterions. The charge ratio shifts to $Z_{-/+} = 0.66$ for the divalent counterions and the brushes returned to the elongated conformation. Scheme 11 describes this switching of the conformation in a schematic way [145].

Simulation were performed to study the influence of counterions of different valencies on the morphology of the brushes [148]. The results are in good agreement with our experimental findings. Moreover, Wegner et al. [146] and Gröhn et al. [149] studied the influence of the aggregating behavior of polyelectrolytes with multivalent counterions.

3.5 *Inter-Polyelectrolyte and Surfactant Complexes of Cylindrical Polyelectrolyte Brushes*

Cylindrical polyelectrolyte brushes can form complexes with oppositely charged surfactants and polyelectrolytes. Although the polyelectrolyte cylindrical brushes behave similarly to their linear analogs when forming complexes with surfactants, their distinctive cylindrical nanostructures make it possible to visualize directly the morphology changes by microscopy such as AFM.

Recently, Schmidt et al. [150] reported helical structures formed by a complex of sodium dodecyl sulfate (SDS) and a cationic polypeptide cylindrical brush. When SDS was in excess, a worm-to-sphere transition was observed. Detailed studies

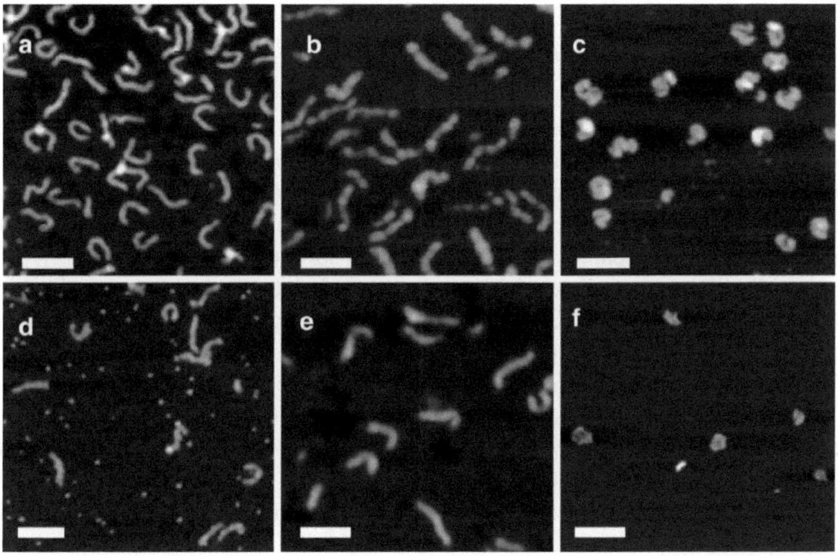

Fig. 17 AFM height images of (**a**) pure PMETAI brushes; (**b**) PMETAI brushes with SDS, with $Z_{-/+} = 0.5$; (**c**) PMETAI brushes with SDS, $Z_{-/+} = 1$; (**d**) PMETAI brushes with SDS, $Z_{-/+} = 1$ with added α-CD (equimolar with SDS); (**e**) same, but with added β-CD; (**f**) sample with β-CD after addition of AdAC (equimolar with β-CD). The *scale bars* represent 200 nm and the brush concentration is $0.02\,\mathrm{g\,L^{-1}}$ [152]. Reprinted by permission of ACS

revealed that the alkyl chain length of the surfactants plays a key role [151]. It is believed that the β-sheet formation of the polypeptide side chains induced the helical conformation of the atactic backbone.

We also investigated the complex formation between SDS and PMETAI cationic brushes[152]. However, no helical intermediate conformations were observed. Only pearl-necklace structures formed at intermediate ratios. When the charge ratio reached unity, the brushes collapsed into spheres. By adding cyclodextrins (α-CD and β-CD), supramolecular inclusion complexes formed with SDS and liberated the brushes. Thus worm-like conformation recovered. A competitive complex agent 1-adamantylammonium chloride (AdAC) can remove the β-CD from the SDS-β-CD complex, causing the recollapse of the brush. This has given some hints to construct sensors or devices on the nano-scale [153]. Figure 17 depicts the conformational transitions of the PMETAI brushes thus effected.

A study of the IPECs formed by polyelectrolyte brushes and oppositely charged linear polyelectrolytes has also been carried out [154, 155]. Recent simulation work indicated that formation of IPECs with linear polyelectrolytes can also induce the collapse of the brushes [156]. Our experimental work [157] concerning the IPECs between PMETAI brushes and linear poly(styrene sulfonate) (PSS) is in agreement with these simulations.

3.6 Amphiphilic and Double-Hydrophilic Core–Shell Cylindrical Brushes

Core–shell cylindrical brushes have gained increasing interest due to their unique properties and applications as nano-templates. Thus, Matyjaszewski et al. and our group first prepared core–shell brushes [30, 134]. We have been studying core–shell brushes containing ionic blocks [30, 31]. Amphiphilic and double hydrophilic core–shell brushes were prepared and studied for their solution properties.

Amphiphilic core–shell brushes in selective solvents can be seen as cylindrical unimolecular micelles [30]. Changing the solvent quality can adjust the thickness of the core or shell of these amphiphilic brushes. Figure 18 presents ^{1}H-NMR spectra of amphiphilic cylindrical brushes with PS-*b*-PAA side chains in different solvents. In CD_3OD, which is a good solvent for PAA but a poor solvent for PS, the signals of the PAA shell are clearly observed, while no signal of PS is detected, indicating the complete collapse of the PS core. By adding $CDCl_3$ to this solution, the PS signals appear in the spectrum (see the range 6.0–7.0 ppm). Now the PS core is solvated by $CDCl_3$ and therefore has an extended conformation.

Simulation work using Monte Carlo (MC) methods have proved that a selective solvent, which is good for the shell but bad for the core, leads to instability in the

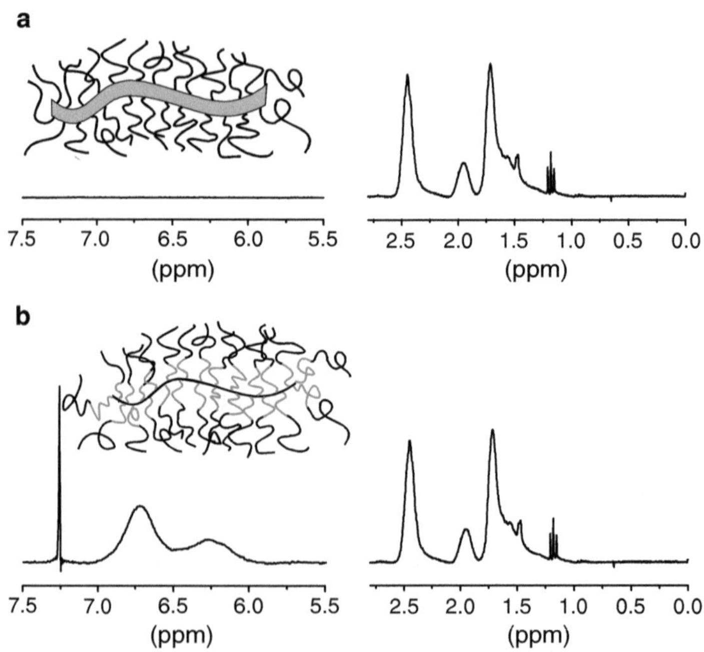

Fig. 18 ^{1}H-NMR spectra of an amphiphilic core–shell polymer brush with PS-b-PAA side chains ($[S_{23}-b-AA_{186}]_{310}$) in (**a**) CD3OD, and (**b**) CD3OD/CDCl3 ($v/v = 1/1$) [30]. Reprinted by permission of ACS

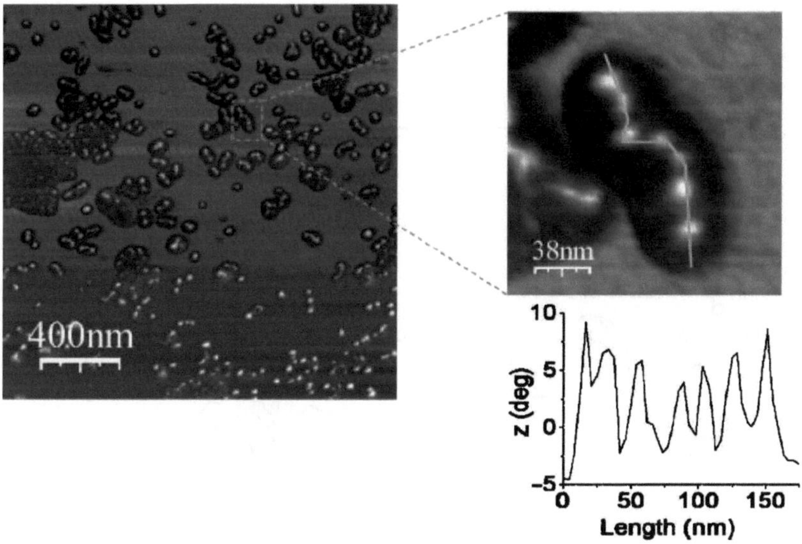

Fig. 19 AFM phase image of amphiphilic brushes $[AA_{25}\text{-}nBA_{61}]_{1500}$ with PAA core, dip coated on mica from toluene solution. The *right image* shows the typical pearl-necklace conformation [158]. Reprinted by permission of ACS

cylindrically uniform structure followed by the appearance of undulations of the collapsed core [158]. Experimental evidence has been reported for this. We have prepared amphiphilic core–shell brushes with PAA as the core and poly(*n*-butyl acrylate) (PnBA) as the shell. In the mixed solvent of methanol and chloroform, they adopt extended worm-like conformations due to the solvation of both the core and the shell. However, in the selective solvent toluene, the PAA core collapsed and so-called pearl-necklace structures formed (see Fig. 19) [158]. The same brushes can also form polychelates with multivalent counterions such as Cd^{2+} and Fe^{3+}. Here similarly undulated structures were also observed [159–161].

We also prepared double hydrophilic core–shell brushes with PMAA as the core and poly(oligoethyleneglycol methacrylate) (POEGMA) as the shell. Due to the weak polyelectrolyte nature of the PMAA, at low pH the brushes also showed typical pearl-necklace conformations, whereas at high pH the core adopts more extended structures [158]. The polychelates of the brush and Fe^{3+} behave in a similar way to the amphiphilic polychelates.

3.7 *One-Dimensional Magnetic Nanohybrid Templated by Core–Shell Brushes*

The core–shell polyelectrolyte brushes can serve as the one-dimensional templates for different nanomaterials since different metal ions or small molecules can be

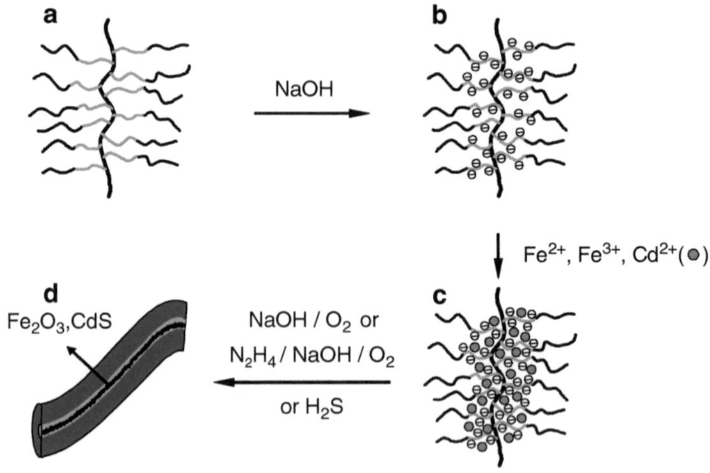

Scheme 12 Schematic illustration for the synthesis of a wire-like assembly of magnetic/semiconducting NPs inside a core–shell cylindrical polymer brush

incorporated inside the brushes. Thus, Schmidt et al. [119, 162] succeeded in the synthesis of a linear array of gold nanoclusters and gold NWs using cylindrical polymer brushes with P2VP core and PS shell as templates. Ishizu et al. [163] prepared polypyrrole within the core of polymer brushes with P2VP-*b*-PMS side chains, aiming to prepare conducting molecular wires. The poly(α-methylstyrene) (PMS) shell of these polymer brushes served as an insulating layer. All these brushes were synthesized by grafting through strategy, and precipitation fractionation was required to obtain brushes from the mixtures with linear MMs.

We have used core–shell cylindrical polymer brushes with PAA core and P*n*BA shell as templates for the growth of different inorganic NWs and nanohybrids, such as CdS [160], CdSe [164], and Fe_2O_3 [161] (see Scheme 12).

Superparamagnetic maghemite (Fe_2O_3) NPs were generated in situ in the core of the PAA-b-PnBA core–shell brushes [31, 161] as described in Scheme 12. The carboxylate groups coordinate with Fe^{3+} ions and form the polychelate. AFM and TEM measurements show distinct pearl-necklace conformations for the polychelates of the brushes with Fe^{3+}. After the alkalization and oxidation reaction small maghemite NPs formed within the core of the brushes distributed in a wire-like manner and the brushes returned to the extended worm-like morphologies.

Figure 20 shows typical magnetization curves. The magnetic properties of the hybrid nanocylinders were investigated at temperatures ranging from 2 to 295 K. It is found that the fabricated NPs are superparamagnetic in the temperature range of 25–295 K, since no hysteresis is observed. At 2 K, magnetization measurement showed a symmetric hysteresis loop, demonstrating the ferrimagnetic nature of the NPs at very low temperatures. The blocking temperature is defined as the temperature above which the particles are free to align with the magnetic field during the measurement time and thus behave superparamagnetically. It is directly related to

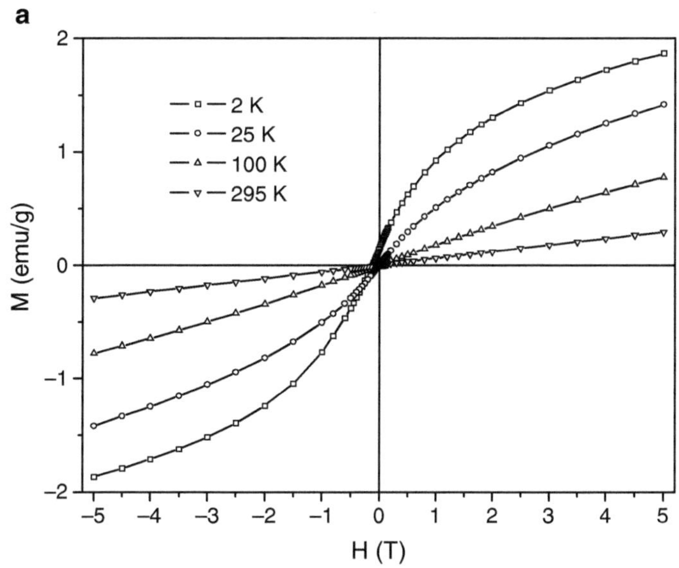

Fig. 20 Magnetization curves for the hybrid nanocylinder of the polymer brush ($[AA_{37}\text{-}nBA_{48}]_{1500}$) and iron oxide nanoparticle (NP) at different temperatures [161]. Reprinted by permission of Wiley-VCH

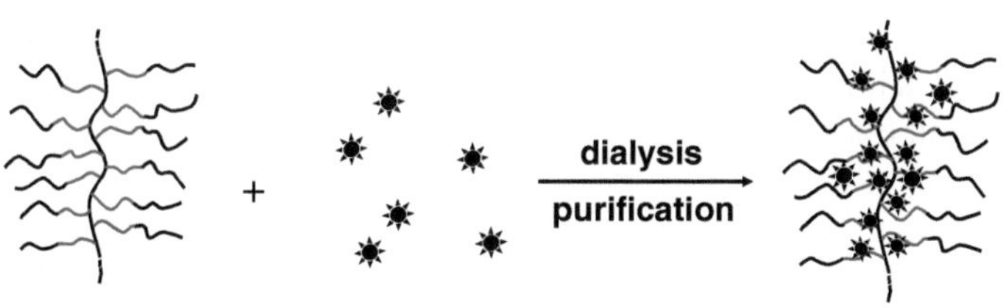

Scheme 13 Introducing-into strategy for the synthesis of magnetic hybrid cylinders

the particle size. For the maghemite particles fabricated in the brushes, the blocking temperature of the hybrid superparamagnetic nanocylinders is extremely low, indicating very small particle sizes. Such small particles have very weak response to the magnetic field. So it is very difficult to manipulate or align these hybrid cylinders at room temperature. Moreover, the solubility of the template amphiphilic core–shell brushes always generates problems in all experiments and mixed organic solvents are needed to disperse the brushes and the hybrids.

In order to improve the solubility and the magnetic response of the hybrid, we have developed a different strategy. First, double hydrophilic core–shell brushes with PMAA as the core and POEGMA as the shell were successfully synthesized [165, 166]. In a separate process, magnetite NPs with size around 10 nm were prepared by coprecipitation. Due to the specific interaction of the carboxylic groups and the magnetite surface, we are able to introduce the magnetite particles into the core of the brushes. This introducing-into strategy is described in Scheme 13. The hybrid also shows superparamagnetic properties at room temperature. Owing to the increased size, we can align the magnetic hybrid in moderate magnetic field on the substrate on a large scale.

4 Conclusions and Outlook

Controlled polymerization techniques have enabled the preparation of well-defined polyelectrolytes of different architectures. Polyelectrolyte stars and cylindrical brushes are two typical examples with isotropic and anisotropic nature, respectively. Different synthetic strategies have been developed for these polyelectrolytes. However, the "core first" and "grafting from" strategies have turned out to be the most suitable methods for the synthesis of polyelectrolyte stars and cylindrical brushes.

Weak polyelectrolyte stars exhibit an interesting behavior in solution, such as higher apparent pK_a for the polyacid stars and higher apparent pK_b for the weak polybase stars, and much lower osmotic pressures, both when comparing them to their linear analogues. Adding monovalent salt reduces the net osmotic pressure and leads to a contraction of the stars. Divalent and trivalent ions cause the collapse of the polyelectrolyte stars. By using a complex trivalent counterion which can undergo a photoaquation reaction, "nanoblossoms" can be built under the UV radiation. The temperature responsiveness of the weak polyelectrolyte PDMAEMA stars was also investigated. Their normal LCST behavior strongly depends on pH. Adding trivalent counterions can even induce a UCST behavior. The stars can also form IPECs with oppositely charged polyelectrolytes.

Polyelectrolyte cylindrical brushes behave in similar ways as the polyelectrolyte stars in many aspects. Due to their anisotropic architecture, their morphologies can be tuned between worms, helices, and spheres. Polyelectrolyte core–shell cylindrical brushes have been used for the fabrication of inorganic NPs or NWs. In particular, superparamagnetic hybrid cylinders with magnetic NPs in the core of the brushes were prepared. They can be aligned on the substrate in magnetic field. This provides another way for the directed assembly of hybrid materials in a controlled manner.

However, there are still challenges remaining in both the synthesis and characterizations of the polyelectrolyte stars and cylindrical brushes. The synthesis still requires multiple steps for the protection–deprotection or derivative reactions. Straightforward methods are needed in the synthesis. The control of the uniformity of the arm number, arm length, side chain number, and side chain length is still a challenge. Nevertheless, more exciting opportunities lie in the study of the polyelectrolyte stars and cylindrical brushes. For example, it is possible to find more ordered phases in the solutions for the stars and brushes. Hence, polyelectrolyte stars and brushes present highly valuable building blocks for nanoscopic and hybrid systems.

Acknowledgments The authors thank Dr. Guanglou Cheng and Dr. Mingfu Zhang for their initial work in the project. Financial support by the Deutsche Forschungsgemeinschaft within the Collaborative Research Center 481 (SFB 481, projects A12 and B9) is gratefully acknowledged.

References

1. Goddard WA III, Brenner DW, Lyshevski SE, Iafrate GJ (eds) (2003) Handbook of nanoscience, engineering, and technology. CRC, Boca Raton
2. Whitesides GM (2005) Small 1:172
3. Gleiter H (1992) Adv Mater 4:474
4. Klabunde KJ (eds) (2001) Nanoscale materials in chemistry. Wiley, New York
5. Schmid G (eds) (2004) Nanoparticles: from theory to application. Wiley-VCH, Weinheim
6. Schwarz JA, Contescu CI, Putyera K (2004) Dekker encyclopedia of nanoscience and nanotechnology-5 volume set. Marcel Dekker, New York
7. Xia Y, Yang P, Sun Y, Wu Y, Mayers B, Gates B, Yin Y, Kim F, Yan H (2003) Adv Mater 15:353
8. Hamley IW (2003) Angew Chem Int Ed 42:1692
9. Bywater S (1979) Adv Polym Sci 30:89
10. Hadjichristidis N (1999) J Polym Sci A Polym Chem 37:857
11. Hatada K, Kitayama T, Ute K, Nishiura T (2004) J Polym Sci A Polym Chem 42:416
12. Inoue K (2000) Prog Polym Sci 25:453
13. Vögtle F, Gestermann S, Hesse R, Schwierz H, Windisch B (2000) Prog Polym Sci 25:987
14. Minko S (2006) Responsive polymer materials: design and applications. Blackwell, Oxford
15. Mandel M (1987) Encycl Polym Sci Eng 11:739
16. Dautzenberg H, Jaeger W, Kötz J, Philipp B, Seidel C, Stscherbina D (1994) Polyelectrolytes: formation, characterization and application. Carl Hanser, Munich
17. Förster S, Abetz V, Müller AHE (2004) Adv Polym Sci: 173
18. Bohrisch J, Eisenbach CD, Jaeger W, Mori H, Müller AHE, Rehahn M, Schaller C, Traser S, Wittmeyer P (2004) Adv Polym Sci 165:1
19. Holm C, Joanny JF, Kremer K, Netz RR, Reineker P, Seidel C, Vilgis TA, Winkler RG (2004) Adv Polym Sci 166:67
20. Dobrynin AV, Rubinstein M (2005) Prog Polym Sci 30:1049
21. Holm C, Rehahn M, Oppermann W, Ballauff M (2004) Adv Polym Sci 166:1
22. Cai Y, Hartenstein M, Müller AHE (2004) Macromolecules 37:7484
23. Mori H, Walther A, André X, Lanzendörfer MG, Müller AHE (2004) Macromolecules 37:2054
24. Mori H, Müller AHE (2003) Prog Polym Sci 28:1403
25. Mori H, Chan Seng D, Lechner H, Zhang M, Müller AHE (2002) Macromolecules 35:9270
26. Erhardt R, Zhang M, Böker A, Zettl H, Abetz C, Frederik P, Krausch G, Abetz V, Müller AHE (2003) J Am Chem Soc 125:3260
27. Plamper FA, Becker H, Lanzendörfer M, Patel M, Wittemann A, Ballauff M, Müller AHE (2005) Macromol Chem Phys 206:1813
28. Plamper FA, Schmalz A, Penott-Chang E, Drechsler M, Jusufi A, Ballauff M, Müller AHE (2007) Macromolecules 40:5689
29. Xu Y, Bolisetty S, Drechsler M, Fang B, Yuan J, Ballauff M, Müller AHE (2008) Polymer 49:3957
30. Cheng G, Böker A, Zhang M, Krausch G, Müller AHE (2001) Macromolecules 34:6883
31. Zhang M, Breiner T, Mori H, Müller AHE (2003) Polymer 44:1449
32. Gao H, Matyjaszewski K (2009) Prog Polym Sci 34:317
33. Sheiko SS, Sumerlin BS, Matyjaszewski K (2008) Prog Polym Sci 33:759
34. Matyjaszewski K, Xia J (2001) Chem Rev 101:2921

35. Barner-Kowollik C, Davis TP, Heuts JPA, Stenzel MH, Vana P, Whittaker M (2003) J Polym Sci A Polym Chem 41:365
36. Blencowe A, Tan JF, Goh TK, Qiao GG (2009) Polymer 50:5
37. Tsitsilianis C, Lutz P, Graff S, Lamps JP, Rempp P (1991) Macromolecules 24:5897
38. Rein D, Rempp P, Lutz PJ (1993) Makromol Chem Macromol Symp 67:237
39. Jacob S, Majoros I, Kennedy JP (1996) Macromolecules 29:8631
40. Kennedy JP, Jacob S (1998) Acc Chem Res 31:835
41. Knischka R, Lutz PJ, Sunder A, Mülhaupt R, Frey H (2000) Macromolecules 33:315
42. Klok H-A, Becker S, Schuch F, Pakula T, Müllen K (2002) Macromol Chem Phys 203:1106
43. Robello DR, Andre A, McCovick TA, Kraus A, Mourey TH (2002) Macromolecules 35:9334
44. Karaky K, Reynaud S, Billon L, Francois J, Chreim Y (2005) J Polym Sci A Polym Chem 43:5186
45. Marsalko TM, Majoros I, Kennedy JP (1993) Polym Bull 31:665
46. Haddleton DM, Crossman MC (1997) Macromol Chem Phys 198:871
47. Rein DH, Rempp P, Lutz P (1998) Macromol Chem Phys 199:569
48. Xia J, Zhang X, Matyjaszewski K (1999) Macromolecules 32:4482
49. Wiltshire JT, Qiao GG (2006) Macromolecules 39:4282
50. Wiltshire JT, Qiao GG (2006) Macromolecules 39:9018
51. Connal LA, Qiao GG (2006) Adv Mater 18:3024
52. Gao H, Matyjaszewski K (2007) J Am Chem Soc 129:11828
53. Gao H, Ohno S, Matyjaszewski K (2006) J Am Chem Soc 128:15111
54. Connal LA, Li Q, Quinn JF, Tjipto E, Caruso F, Qiao GG (2008) Macromolecules 41:2620
55. Kim B-S, Gao H, Argun AA, Matyjaszewski K, Hammond PT (2009) Macromolecules 42:368
56. Furukawa T, Uchida S, Ishizu K (2007) J Appl Polym Sci 105:1543
57. Held D, Müller AHE (2000) Macromol Symp 157:225
58. Angot S, Murthy KS, Taton D, Gnanou Y (1998) Macromolecules 31:7218
59. Angot S, Murthy KS, Taton D, Gnanou Y (2000) Macromolecules 33:7261
60. Taton D, Saule M, Logan J, Duran R, Hou S, Chaikof EL, Gnanou Y (2003) J Polym Sci A Polym Chem 41:1669
61. Kudo H, Inoue H, Inagaki T, Nishikubo T (2009) Macromolecules 42:1051
62. Ohno K, Wong B, Haddleton DM (2001) J Polym Sci A Polym Chem 39:2206
63. Stenzel-Rosenbaum MH, Davis TP, Chen V, Fane AG (2001) Macromolecules 34:5433
64. Narumi A, Satoh T, Kaga H, Kakuchi T (2002) Macromolecules 35:699
65. Kaga H, Yamane S, Narumi A, Satoh T, Kakuchi T (2004) Macromol Symp 217:29
66. Stenzel MH, Davis TP (2002) J Polym Sci A Polym Chem 40:4498
67. Miura Y, Narumi A, Matsuya S, Satoh T, Duan Q, Kaga H, Kakuchi T (2005) J Polym Sci A Polym Chem 43:4271
68. Li J, Xiao H, Kim YS, Lowe TL (2005) J Polym Sci A Polym Chem 43:6345
69. Schumacher M, Ruppel M, Kohlbrecher J, Burkhardt M, Plamper F, Drechsler M, Müller AHE (2009) Polymer 50:1908
70. Gozgen A, Dag A, Durmaz H, Sirkecioglu O, Hizal G, Tunca U (2009) J Polym Sci A Polym Chem 47:497
71. Barner L, Barner-Kowollik C, Davis TP, Stenzel MH (2004) Aust J Chem 57:19
72. Hong C-Y, You Y-Z, Liu J, Pan C-Y (2005) J Polym Sci A Polym Chem 43:6379
73. Schnitter M, Engelking J, Heise A, Miller RD, Menzel H (2000) Macromol Chem Phys 201:1504
74. Strandman S, Hietala S, Aseyev V, Koli B, Butcher SJ, Tenhu H (2006) Polymer 47:6524
75. Shen Z, Chen Y, Barriau E, Frey H (2006) Macromol Chem Phys 207:57
76. Strandman S, Zarembo A, Darinskii AA, Loeflund B, Butcher SJ, Tenhu H (2007) Polymer 48:7008
77. Strandman S, Zarembo A, Darinskii AA, Laurinmaki P, Butcher SJ, Vuorimaa E, Lemmetyinen H, Tenhu H (2008) Macromolecules 41:8855
78. Shim Y-H, Bougard F, Coulembier O, Lazzaroni R, Dubois P (2008) Eur Polym J 44:3715
79. Plamper FA, Walther A, Müller AHE, Ballauff M (2007) Nano Lett 7:167

80. Plamper FA, Ruppel M, Schmalz A, Borisov O, Ballauff M, Müller AHE (2007) Macromolecules 40:8361
81. Plamper FA, Schmalz A, Ballauff M, Müller AHE (2007) J Am Chem Soc 129:14538
82. Pergushov DV, Babin IA, Plamper FA, Zezin AB, Müller AHE (2008) Langmuir 24:6414
83. Muthukrishnan S, Plamper F, Mori H, Müller AHE (2005) Macromolecules 38:10631
84. Manning GS (1969) J Chem Phys 51:924
85. Borisov OV, Zhulina EB (1998) Eur Phys J B 4:205
86. Plamper FA, Becker H, Lanzendörfer M, Patel M, Wittemann A, Ballauff M, Müller AHE (2005) Macromol Chem Phys 206:1813
87. Pincus P (1991) Macromolecules 24:2912
88. Jusufi A, Likos CN, Löwen H (2002) Phys Rev Lett 88:018301
89. Jusufi A, Likos CN, Löwen H (2002) J Chem Phys 116:11011
90. Deserno M, Holm C, Blaul J, Ballauff M, Rehahn M (2001) Eur Phys J E 5:97
91. Mei Y, Ballauff M (2005) Eur Phys J E 16:341
92. Mei Y, Lauterbach K, Hoffmann M, Borisov OV, Ballauff M, Jusufi A (2006) Phys Rev Lett 97:158301/1
93. Likos CN, Hoffmann N, Jusufi A, Lowen H (2003) J Phys Condens Matter 15:S233
94. Hoffmann N, Likos CN, Löwen H (2004) J Chem Phys 121
95. Furukawa T, Ishizu K (2005) Macromolecules 38:2911
96. Kabanov VA, Kargina OV, Ulyanova MV (1976) J Polym Sci Polym Chem Ed 14:2351
97. Kabanov VA, Kargina OV, Mishustina LA, Lubanov SY, Katuzynski K, Penczek S (1981) Makromol Chem Rapid Commun 2:343
98. Izumrudov VA, Savitskii AP, Bakeev KN, Zezin B, Kabanov VA (1984) Makromol Chem Rapid Commun 5:709
99. Kabanov VA, Zezin AB (1984) Macromol Chem Phys Suppl 6:259
100. Kabanov VA, Zezin AB (1984) Pure Appl Chem 56:343
101. Izumrudov VA, Bronich TK, Zezin AB, Kabanov VA (1985) J Polym Sci Polym Lett Ed 23:439
102. Yin M, Ding K, Gropeanu RA, Shen J, Berger R, Weil T, Mullen K (2008) Biomacromolecules 9:3231
103. Ding J, Wang L, Yu H, Huo J, Liu Q, Xiao A (2009) J Phys Chem C 113:5126
104. Li J, Shi L, An Y, Li Y, Chen X, Dong H (2006) Polymer 47:8480
105. Georgiades SN, Vamvakaki M, Patrickios CS (2002) Macromolecules 35:4903
106. Mengel C, Meyer WH, Wegner G (2001) Macromol Chem Phys 202:1138
107. Chu C-C, Wang Y-W, Wang L, Ho T-I (2005) Synth Met 153:321
108. Tsukahara Y, Mizuno K, Segawa A, Yamashita Y (1989) Macromolecules 22:1546
109. Dziezok P, Sheiko SS, Fischer K, Schmidt M, Möller M (1997) Angew Chem Int Ed 36:2812
110. Pyun J, Kowalewski T, Matyjaszewski K (2003) Macromol Rapid Commun 24:1043
111. Zhang M, Müller AHE (2005) J Polym Sci A Polym Chem 43:3461
112. Frey H (1998) Angew Chem Int Ed 37:2193
113. Schlüter AD (1998) Top Curr Chem 197:165
114. Schluter AD, Rabe JP (2000) Angew Chem Int Ed 39:864
115. Zhang A, Shu L, Bo Z, Schluter AD (2003) Macromol Chem Phys 204:328
116. Frauenrath H (2005) Prog Polym Sci 30:325
117. Barriau E, Marcos AG, Kautz H, Frey H (2005) Macromol Rapid Commun 26:862
118. Marcos AG, Pusel TM, Thomann R, Pakula T, Okrasa L, Geppert S, Gronski W, Frey H (2006) Macromolecules 39:971
119. Djalali R, Hugenberg N, Fischer K, Schmidt M (1999) Macromol Rapid Commun 20:444
120. Neiser MW, Okuda J, Schmidt M (2003) Macromolecules 36:5437
121. Zhang B, Fischer K, Schmidt M (2005) Macromol Chem Phys 206:157
122. Ishizu K, Shen XX (1999) Polymer 40:3251
123. Ishizu K, Shen XX, Tsubaki KI (2000) Polymer 41:2053
124. Ishizu K (2004) Polym J 36:775
125. Ishizu K, Satoh J, Toyoda K, Sogabe A (2004) J Mater Sci 39:4295
126. Ishizu K, Yamada H (2007) Macromolecules 40:3056

127. Ishizu K, Takano S, Ochi K (2007) J Appl Polym Sci 104:3994
128. Rühe J, Ballauff M, Biesalski M, Dziezok P, Gröhn F, Johannsmann D, Houbenov N, Hugenberg N, Konradi R, Minko S, Motornov M, Netz RR, Schmidt M, Seidel C, Stamm M, Stephan T, Usov D, Zhang H (2004) Adv Polym Sci 165:79
129. Hua F, Kita R, Wegner G, Meyer W (2005) Chemphyschem 6:336
130. Deffieux A, Schappacher M (1999) Macromolecules 32:1797
131. Ryu SW, Hirao A (2000) Macromolecules 33:4765
132. Gao H, Matyjaszewski K (2007) J Am Chem Soc 129:6633
133. Beers KL, Gaynor SG, Matyjaszewski K, Sheiko SS, Möller M (1998) Macromolecules 31:9413
134. Börner HG, Beers K, Matyjaszewski K, Sheiko SS, Möller M (2001) Macromolecules 34:4375
135. Qin S, Börner HG, Matyjaszewski K, Sheiko SS (2002) Polymer Prepr 43:237
136. Yuan J, Xu Y, Walther A, Bolisetty S, Schumacher M, Schmalz H, Ballauff M, Müller AHE (2008) Nat Mater 7:718
137. Lee H, Jakubowski W, Matyjaszewski K, Yu S, Sheiko SS (2006) Macromolecules 39:4983
138. Cheng C, Qi K, Khoshdel E, Wooley KL (2006) J Am Chem Soc 128:6808
139. Neugebauer D, Sumerlin BS, Matyjaszewski K, Goodhart B, Sheiko SS (2004) Polymer 45:8173
140. Sumerlin BS, Neugebauer D, Matyjaszewski K (2005) Macromolecules 38:702
141. Xu Y, Becker H, Yuan J, Burkhardt M, Zhang Y, Walther A, Bolisetty S, Ballauff M, Müller AHE (2007) Macromol Chem Phys 208:1666
142. Lienkamp K, Noe L, Breniaux M-H, Lieberwirth I, Gröhn F, Wegner G (2007) Macromolecules 40:2486
143. Lienkamp K, Ruthard C, Lieser G, Berger R, Groehn F, Wegner G (2006) Macromol Chem Phys 207:2050
144. Xu Y, Walther A, Müller AHE (2009) submitted
145. Xu Y, Bolisetty S, Drechsler M, Fang B, Yuan J, Harnau L, Ballauff M, Müller AHE (2009) Soft Matter 5:379
146. Kroeger A, Belack J, Larsen A, Fytas G, Wegner G (2006) Macromolecules 39:7098
147. Snir Y, Kamien RD (2005) Science 307:1067
148. Yan L-T, Xu Y, Ballauff M, Müller AHE, Boker A (2009) J Phys Chem B 113:5104
149. Ruthard C, Maskos M, Kolb U, Groehn F (2009) Macromolecules 42:830
150. Gunari N, Cong Y, Zhang B, Fischer K, Janshoff A, Schmidt M (2008) Macromol Rapid Commun 29:821
151. Cong Y, Gunari N, Zhang B, Janshoff A, Schmidt M (2009) Langmuir:DOI: 10.1021/-la804290r
152. Xu Y, Bolisetty S, Ballauff M, Müller AHE (2009) J Am Chem Soc 131:1640
153. Dai L (2003) Intelligent macromolecules for smart devices: from materials synthesis to device applications. Springer, London
154. Stoerkle D, Duschner S, Heimann N, Maskos M, Schmidt M (2007) Macromolecules 40:7998
155. Duschner S, Stoerkle D, Schmidt M, Maskos M (2008) Macromolecules 41:9067
156. Yan L-T, Zhang X (2009) Langmuir 25:3808
157. Xu Y, Müller AHE (2009) submitted
158. Polotsky A, Charlaganov M, Xu Y, Leermakers FAM, Daoud M, Müller AHE, Dotera T, Borisov O (2008) Macromolecules 41:4020
159. Zhang M, Teissier P, Krekhova M, Cabuil V, Müller AHE (2004) Prog Colloid Polym Sci 126:35
160. Zhang M, Drechsler M, Müller AHE (2004) Chem Mater 16:537
161. Zhang M, Estournes C, Bietsch W, Müller AHE (2004) Adv Funct Mater 14:871
162. Djalali R, Li S-Y, Schmidt M (2002) Macromolecules 35:4282
163. Ishizu K, Tsubaki K, Uchida S (2002) Macromolecules 35:10193
164. Yuan J, Drechsler M, Xu Y, Zhang M, Müller AHE (2008) Polymer 49:1547
165. Xu Y, Drechsler M, Yuan J, Müller AHE (2008) Polymer Prepr 49(1):338
166. Xu Y, Drechsler M, Yuan J, Müller AHE, submitted

Adv Polym Sci (2010) 228: 39–58
DOI:10.1007/12_2010_52

Published online: 25 March 2010

Various Aspects of the Interfacial Self-Assembly of Nanoparticles

Nicole Popp, Sergej Kutuzov, and Alexander Böker

Abstract We describe the interfacial self-assembly of nanoparticles at liquid–liquid interfaces and in block copolymers. At the interface of two immiscible liquids, the particles assemble into disordered but densely packed monolayers. This self-assembly process was investigated ex situ with scanning force microscopy (SFM) and transmission electron microscopy (TEM), and laser scanning confocal microscopy (LCSM) methods. Adsorbed particles can be crosslinked at the interface to fabricate mechanically stable capsules and membranes. In addition, it was shown by pendant drop tensiometry that Janus particles consisting of gold and iron oxide show a significantly higher interfacial activity than homogeneous gold or iron oxide nanoparticles of comparable size and chemical nature. For the self-assembly of nanoparticles in block copolymer mixtures, it was shown theoretically and experimentally that these composite materials form hierarchically ordered structures. Therefore, thin films from mixtures of a cylindrical polystyrene-*block*-poly(2-vinylpyridine), with tri-*n*-octylphosphine oxide-covered CdSe nanoparticles were prepared and investigated with SFM, TEM, and grazing-incidence small angle x-ray scattering (GISAXS) after thermal annealing.

Keywords Block copolymer/nanoparticle mixtures · Interfacial assembly · Nanoparticles · Janus particles

---

N. Popp and S. Kutuzov
Lehrstuhl für Physikalische Chemie II, Universität Bayreuth, 95440 Bayreuth, Germany
e-mail: nicole.popp@uni-bayreuth.de; Sergey.Kutuzov@uni-bayreuth.de

A. Böker (✉)
Lehrstuhl für Makromolekulare Materialien und Oberflächen and DWI an der RWTH Aachen e.V., RWTH Aachen University, 52056 Aachen, Germany
and
Lehrstuhl für Physikalische Chemie II, Universität Bayreuth, 95440 Bayreuth, Germany
e-mail: boeker@dwi.rwth-aachen.de

Contents

Abbreviations

DFT	Density functional theory
FLIP	Fluorescence loss induced by photobleaching
FRAP	Fluorescence recovery after photobleaching
GISAXS	Grazing-incidence small-angle X-ray scattering
LbL	Layer-by-layer
O	Oil
PS	Polystyrene
PS-*b*-P2VP	Polystyrene-*block*-poly(2-vinyl-piridine)
RhB	Rhodamine B
SCFT	Self-consistent field theory
SFM	Scanning force microscopy
TEM	Transmission electron microscopy
TOPO	Tri-*n*-octylphosphine oxide
W	Water

1 Introduction

This chapter is concerned with the increasing use of liquid interfaces as templates for the self-assembly of colloidal particles [1]. In particular, besides micrometer-sized colloids, nanoscopic particles with sizes down to a few nanometers have recently been investigated. On one hand, control of the structure formation processes at the nanometer level poses a challenging problem. On the other hand, the use of nanoparticles yields ample opportunities for the fabrication of nanostructured devices (e.g., nanoporous containers or filtering devices). In addition to the use of classical oil–water (O/W) emulsion systems, like the so-called Pickering emulsions, fluid interfaces such as found in block copolymer nanostructures can be employed. Here, the nanoparticles impart specific functions to the nanostructures, such as magnetism or charge transport, as required in magnetic data storage media or polymer-based photovoltaic devices, respectively.

1.1 Pickering Emulsions

More than a century ago, Pickering [2] and Ramsden [3] investigated paraffin–water emulsions contains solid particles such as iron oxide, silicon dioxide, barium sulfate, and kaolin and discovered that these micron-sized colloids generate a resistant film at the interface between the two immiscible phases, inhibiting the coalescence of the emulsion drops. These so-called Pickering emulsions are formed by the self-assembly of colloidal particles at fluid–fluid interfaces in two-phase liquid systems (Fig. 1).

The desorption energy, which is directly related to the stability of emulsions, depends on the particle size, particle–particle interaction and, of course, particle–water and particle–oil interactions [4, 5].

Almost 80 years after Pickering's discovery, the behavior of the colloidal particles was described theoretically by Pieranski, who argued that the assembly of spherical particles at the O/W interface was determined by a decrease of the total free energy [5]. The placement of a single particle with an effective radius r, at the interface between an oil and water leads to a decrease of the initial interfacial energy E_0 to E_1 yielding an energy difference of ΔE_1:

$$E_1 - E_0 = \Delta E_1 = -\frac{\pi \cdot r^2}{\gamma_{O/W}} \cdot \left[\gamma_{O/W} - (\gamma_{P/W} - \gamma_{P/O})\right]^2 \tag{1}$$

Here, the three contributions to the interfacial energy arise from the particle–oil interface ($\gamma_{P/O}$), the particle–water interface ($\gamma_{P/W}$), and the oil–water interface ($\gamma_{O/W}$). From (1) it is evident that, for a given emulsion system (i.e., with fixed $\gamma_{P/O}$, $\gamma_{P/W}$, and $\gamma_{O/W}$), the stability of the particle assembly is determined by the square of the particle radius r. For microscopic particles the decrease in total free energy is much larger than the thermal energy (a few k_BT, where k_B is the Boltzmann constant and T is temperature) leading to an effective confinement of large colloids to the interface. Nanoscopic particles, however, are confined to the interface by an energy reduction comparable to thermal energy. Consequently, nanoparticles are

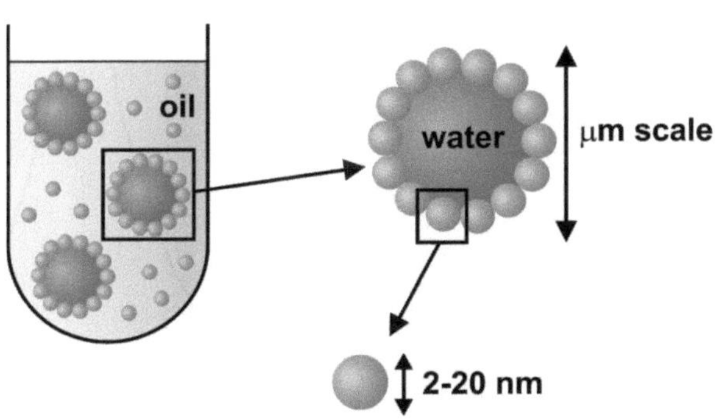

Fig. 1 Self-assembly of solid nanoparticles at the oil–water interface

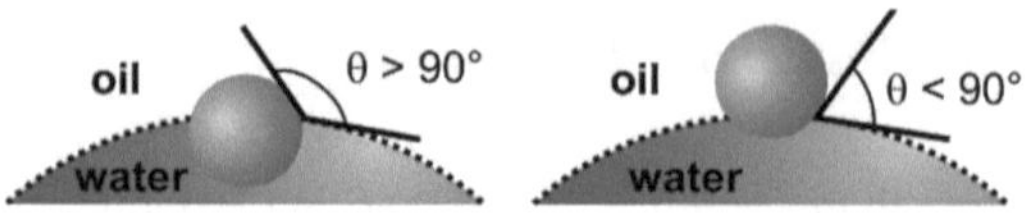

Fig. 2 Changes in wettability of solid particles at the oil–water interface at contact angles $\theta > 90°$ and $< 90°$

easily displaced from the interface, leading to a constant particle exchange at the interface, the rate of which depends on particle size. The thermally activated escape of small particles takes place more often than for larger ones and, for the equilibrium state of assembly, the total gain in free energy is lower for smaller particles. Therefore, assemblies of larger particles are more stable. This size dependence allows the nanoparticle assembly to attain their equilibrium structure at the interface, whereas micrometer-size colloids might be trapped in a non-equilibrium state. Following these theoretical thoughts, various colloidal systems with particles of different size and surface chemistry (e.g., polystyrene lattices, silica particles, etc.) were described in the literature [4, 6–9]. Moreover, the behavior of nanometer-size particles was investigated in detail due to their high potential for the construction of hierarchical and functional structures [10–13].

In addition to the size of the nanoparticles, the interfacial tension and, therefore, the wettability of a particle surface, also dictates the desorption energy [4]. The wettability is described by the contact angle θ between the solid and the oil–water interface. The stability of oil-in-water (O/W) emulsions or water-in-oil (W/O) emulsions depends on this contact angle. In general, the less-wetting liquid becomes the dispersed phase. If θ is lower than 90°, O/W emulsions are more stable; at contact angles greater than 90°, W/O emulsions are favored (Fig. 2) [14].

The variation of the desorption energy with the contact angle is displayed in Fig. 3. Binks and Lumsdon investigated a toluene–water system with constant interfacial tension of 36 mN/m by using silica nanoparticles of constant radius of 10 nm and various wettabilities [4]. At a contact angle of 90°, a maximum in desorption

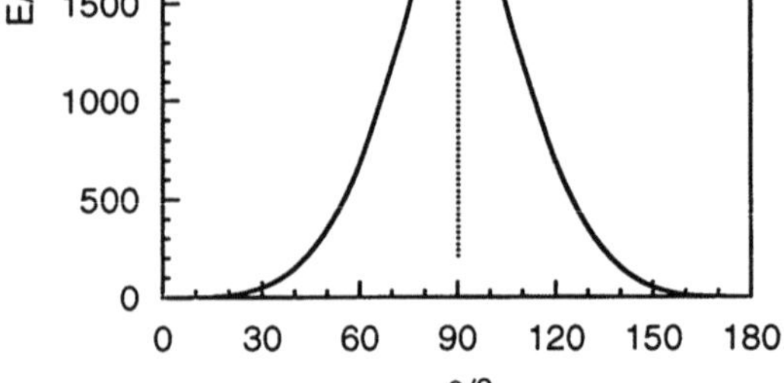

Fig. 3 Variation of the desorption energy E (equal to k_BT) of a spherical particle at a planar oil–water interface as a function of the contact angle θ. The depth of immersion into water is shown as h, the particle radius (R) is 10 nm, and the interfacial tension ($\gamma_{O/W}$) is 36 mN/m. Reprinted with permission from Langmuir [4]. Copyright (2000) American Chemical Society

energy was observed. Increasing or decreasing the contact angle, starting from the maximum stabilization at 90°, also decreased the stability of the emulsion. If the contact angle was between 0 and 20° or between 160 and 180°, the energy was $10\,k_BT$ or less, which is on the order of magnitude of the desorption energy of conventional surfactants [15].

So far, the stabilizing effect has only been described for particles with a homogeneous surface, i.e., a surface with homogeneous wettability. Later in this chapter, we will give a brief introduction to the theoretically anticipated behavior of particles with a heterogeneous surface exhibiting heterogeneous wettability.

1.2 Nanoparticles as Building Blocks

Besides the basic interest in the parameters governing particle interfacial assembly, there is also considerable technological potential associated with the structures formed at liquid–liquid interfaces. For example, nanoparticles could serve as building blocks for capsules and membranes with nanoscopic pores for filtering or encapsulation and for delivery purposes.

During recent years, several approaches for the design of nano- and microscopic capsules have been described in the literature. The electrostatic adsorption of polyelectrolytes [16–18] or particles [19] has also been investigated. Capsules have been successfully produced by polymerization in so-called mini-emulsions [20]. The most promising process so far uses a nanometer-to-micrometer-sized solid template of a water-insoluble substance, which is subsequently degraded to yield a hollow material [21–23]. These capsules swell in appropriate solvents and are filled by a diffusion-controlled process from the surrounding phase. The drawbacks of this method are the broad pore size distribution, and the ex situ filling procedure, during which only substances sufficiently small to pass the pores from the outside can be inserted into the capsule. Further experiments have focused on the use of liquid–liquid interfaces (i.e. water droplets dispersed in oil or a flat oil–water interface) as templates for the production of microporous capsules and membranes. Schüth, Fowler and others grew zeolite structures or silica spheres at such interfaces, but failed to generate structures with defined pore sizes and size distribution [24–28].

Pore size control in microparticle colloidal assemblies was reported by Dinsmore et al., using assembly of the particles at fluid interfaces, followed by sintering. The resulting interstices of the quasi-hexagonal array of particles gave pores in the range of several hundred nanometers [29, 30]. Furthermore, Gödel et al. fabricated elastic, nanoporous membranes from silica-polyisoprene hybrid materials spread in a Langmuir trough. Crosslinking the nanoparticle-polyisoprene film by UV radiation, followed by dissolution of the silica particles gave robust polymer membranes with pore sizes between 30 and 500 nm [31–34].

It remains a challenge, however, to fabricate capsules and membranes with precisely controlled pore size and a pore size distribution on the lower nanometer scale. Capsules with pores between 5 and 20 nm have been a long-standing goal for encapsulation and immuno-isolation of cells for treatment of diabetes, cancer and other illnesses [35–38]. After implantation of living cells into a host, pores of this size protect the cells from the host's immune response, yet allow exchange of nutrients and secreted chemicals. Existing approaches towards fabrication of immuno-isolating capsules might result in a distribution of pore sizes too broad to be effective [35], or might require laborious lithographic processing of one capsule at a time [39].

A novel approach for the synthesis of such materials lies in the use of a nanoparticle toolkit that consists of particles with diameters ranging from 2 to 10 nm. The necessary toolkit is readily available due to the developments on nanoparticle synthesis (CdSe, Au, SiO_x, etc.) and surface functionalization methods using state-of-the-art techniques [40, 41]. So far, there are only a few publications dealing with the fluid-directed self-assembly of nanoparticles. Many papers, however, describe particle self-assembly at solid interfaces. Korgel et al. for example created self-organized superstructures from 5 nm gold particles [42–44]. Analogous experiments using CdSe, CdTe, and HgTe nanoparticles were described by Bawendi et al. [45]. First studies on the nanoparticle self-assembly at curved (droplet) interfaces in Pickering emulsions were recently published by Lin et al. [12, 46, 47], Mann et al. [48], Dai et al. [49], and Möhwald and coworkers [10, 11, 13].

1.3 Homogeneous Nanoparticles at Fluid Interfaces

The structure of CdSe nanoparticles segregated to the fluid interface as shown by confocal microscopy (Fig. 4a) has been investigated ex situ with scanning force microscopy (SFM) and transmission electron microscopy (TEM) methods. All results point to a monolayer of nanoparticles with liquidlike ordering at the interface (Fig. 4b,c) [46].

In another approach, the interfacial diffusion of the nanoparticles was determined using two photobleaching methods: fluorescence loss induced by photobleaching (FLIP) and fluorescence recovery after photobleaching (FRAP). It was found that the lateral diffusion of the nanoparticles at the interface as well as the diffusion normal to and from the interface deviated by about four orders of magnitude from the values obtained in free solution [46].

Moreover, a study using pendant drop tensiometry to follow the change in interfacial tension accompanied by simultaneous ex situ TEM measurements yielded insight into the mechanism of nanoparticle adsorption to the liquid interface [50]. As can be inferred from the data in Fig. 5, different stages of adsorption can be distinguished. The TEM images in Fig. 6 show the mechanism of nanoparticle monolayer formation in detail. First, free nanoparticle diffusion to the interface occurs. Secondly, the particles pack closer and form clusters that grow to form a closely packed particle array, lowering the interfacial tension. Finally, thermally activated exchange between adsorbed and incoming particles is observed leading to a tightly packed monolayer and only a slow decrease in interfacial tension at later times.

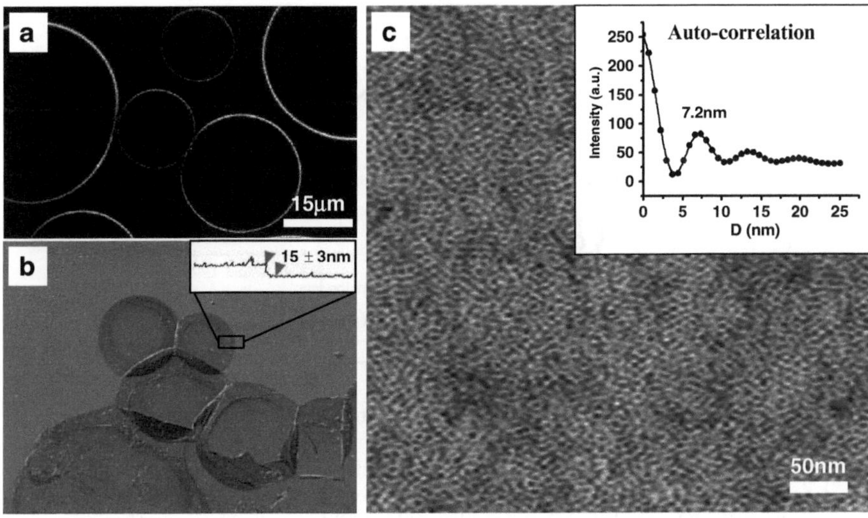

Fig. 4 (**a**) Fluorescence confocal microscope image of water droplets dispersed in toluene, covered with CdSe nanoparticles. (**b**) Differential interference contrast optical microscopy image of dried droplets on a silicon substrate. *Inset*: The atomic force microscopy height section analysis shows twice the thickness of a monolayer. (**c**) TEM image of a dried droplet. *Inset*: Auto-correlation function of the TEM image reveals a mean particle distance of 7.2 nm, which is in good agreement with the value expected for particles with 4.6 nm diameter and 0.8 nm hydrocarbon ligand. Reprinted with permission from Langmuir [46]. Copyright (2005) American Chemical Society

These observations point to the formation of a nanoparticle monolayer by nucleation and growth. Furthermore, the relationship between the free diffusion and the diffusion for the late stage of adsorption as calculated from the changes in interfacial tension reveals an energy barrier at late stages that corresponds to the activation energy for a thermally triggered escape of nanoparticles from the interface. This is in good agreement with the observed packing behavior.

To fabricate mechanically stable capsules and membranes from the spherical nanoparticle assemblies, the adsorbed particles need to be crosslinked at the interface. This can be done by the use of nanoparticles that are stabilized by reactive organic molecules. CdSe nanoparticles, stabilized by benzene vinyl ligands, segregated to the toluene–water interface and were subsequently crosslinked using a water-soluble radical initiator. This process yielded robust membranes that maintained their integrity even when removed from the interface (Fig. 7) [47]. Such crosslinked nanoparticle assemblies show high elasticity, extraordinary stability in water, and even serve as effective diffusion barriers for small molecule dyes. Using the process described above, a nanoparticle membrane has been generated in an Eppendorf tube at the toluene–water interface (Fig. 8). After removal of the organic phase and introduction of an aqueous Rhodamine B solution, a well-defined diffusion of the dye across the membrane was observed within about 15 min, without any sign of turbulent mixing (Fig. 8) [47].

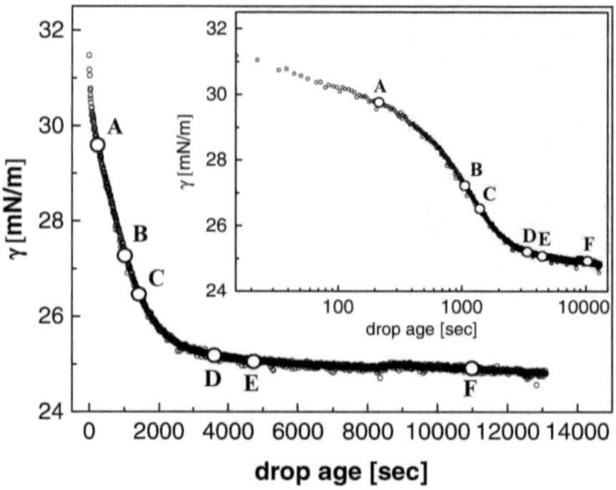

Fig. 5 Dynamic interfacial tension (γ) measurements of a toluene–water interface during adsorption of 6-nm CdSe nanoparticles to a pendant water drop in toluene (CdSe concentration was 1.58×10^{-6} mol/L). The *circles* mark the time at which TEM samples shown in Fig. 6 were prepared. The *inset* depicts the data on a logarithmic time scale. Reprinted with permission from Physical Chemistry Chemical Physics [50]. Copyright (2007) RSC Publishing

Consequently, the use of functionalized ligands attached to the nanoparticles is shown to provide an effective means of stabilizing the interfacial assembly by crosslinking. Moreover, the nanoparticle assembly proved to be as elastic and robust

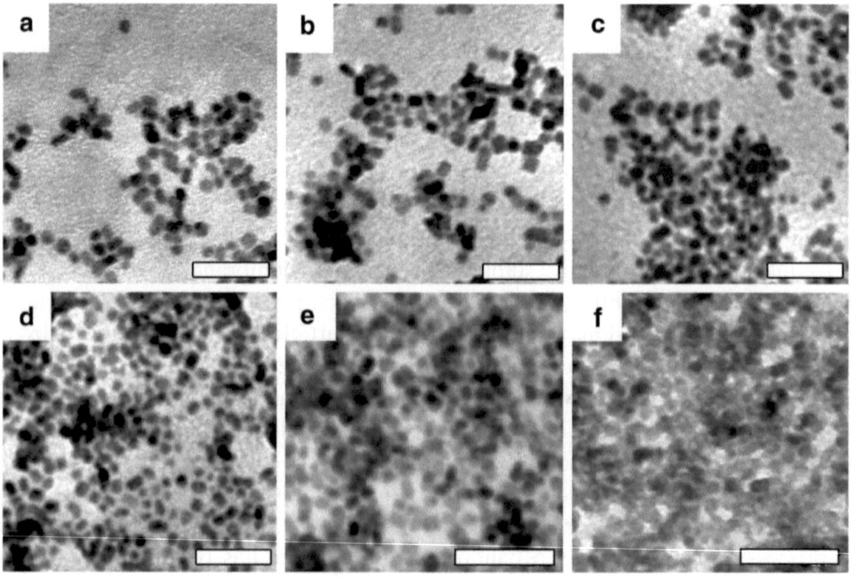

Fig. 6 Series of TEM images of 6-nm nanoparticle adsorption to the toluene–water interface at different adsorption times as marked in Fig. 5: (**a**) 230 s, (**b**) 1060 s, (**c**) 1400 s, (**d**) 3514 s, (**e**) 4700 s, and (**f**) 10,800 s. Structure formation via nucleation and growth of clusters can be seen. Scale bars: 40 nm. Reprinted with permission from Physical Chemistry Chemical Physics [50]. Copyright (2007) RSC Publishing

Fig. 7 Confocal microscope image from different viewing angles of a nanoparticle sheet prepared by crosslinking the functional ligands. Scale bars: 50 μm. Reprinted with permission from Journal of the American Chemical Society [47]. Copyright (2003) American Chemical Society

as expected for a nanometer thin sheet of polystyrene (PS). This work can be seen as a proof-of-concept for the use of nanoparticles as building blocks for nanoporous membranes and capsules.

A drawback of the crosslinking strategy used initially for nanoparticle assemblies is the elevated temperature (~60°C) necessary to initiate the radical crosslinking of the vinyl benzene ligands Ligand systems have since been developed that allow crosslinking at room temperature. Ring-opening metathesis polymerization was employed using a norbornene derivative as the ligand attached to CdSe/ZnS core–shell nanoparticles in combination with a water-soluble poly(ethylene glycol)-conjugated (PEGylated) Grubbs catalyst. It was shown that these novel nanoparticles form stable assemblies at the toluene–water interface and that these assemblies could be crosslinked to yield well-defined CdSe/ZnS capsules [51]. Recently, Kotov et al. reported on the spontaneous formation of two-dimensional free-floating nanoparticle sheets from tetrahedral CdTe nanocrystals in solution [52]. Computer simulations revealed the interplay between the

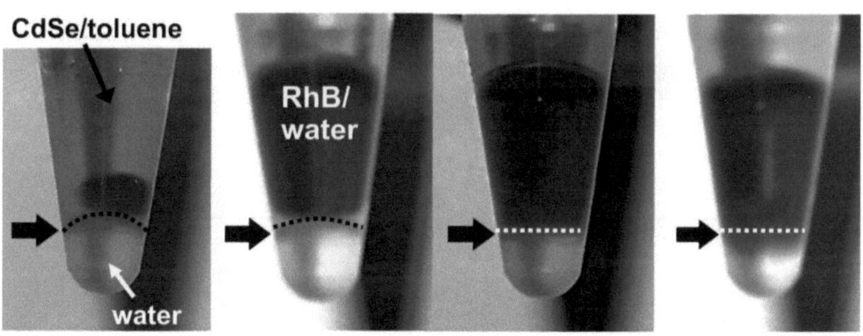

Fig. 8 Rhodamine B dye (RhB, red solution) diffusing across a membrane of crosslinked nanoparticles (*dotted line*). The *bold arrows* point to the interface in each tube. The two *right-hand* images represent a time frame of about 15 min. Subsequent addition of water to the RhB/water droplet replaces the CdSe/toluene solution leading to a RhB/water–water interface separated by the nanoparticle membrane. Reprinted with permission from Journal of the American Chemical Society [47]. Copyright (2003) American Chemical Society

electrostatic interaction and the anisotropic hydrophobic attraction between the nanoparticles to be crucial for the aggregation process. In addition, Tsukruk et al. generated free-standing polymer-nanoparticle composite films of only 20 nm thickness for sensor applications using a layer-by-layer (LbL) assembly technique [53].

1.4 Janus Particles at Fluid Interfaces

A Janus particle is defined as having two distinctly different hemispherical or surface regions; polar and apolar regions are one example. Such a particle is characterized by the contact angle of the polar region and the contact angle of the apolar region. The ratio of the areas of the polar and apolar regions can be changed, where the angle α specifies the position of the surface boundary between both regions. (Fig. 9) Values of α of either 0° or 180° correspond to homogeneous particles, while a Janus particle in the original meaning would have a value of α of 90° due to the equal polar and apolar regions [54].

As discussed by Binks and Lumsdon, amphiphilic Janus particles can exhibit an interfacial activity several times higher than simple homogeneous particles [54]. Janus particles combine the amphiphilic character of surfactants and the physical properties of nanoparticles, which opens new opportunities in emerging areas of nanotechnology and emulsion stabilization.

Recently, Perro et al. [55] reviewed the developments in the field of Janus particles over the last 15 years, describing various strategies to obtain Janus-type particles using polymer precursors. One strategy is based on the self-assembly of ABC terpolymers in bulk [56, 57] or in solution [58]. Another uses the electrostatic interactions of AB and CD diblock copolymers, which lead to inter-polyelectrolyte complexes [59]. A different synthetic concept is to obtain Janus particles made of inorganic materials, e.g., acorn-like particles made of PdS_x–Co_9S_8 [60] or

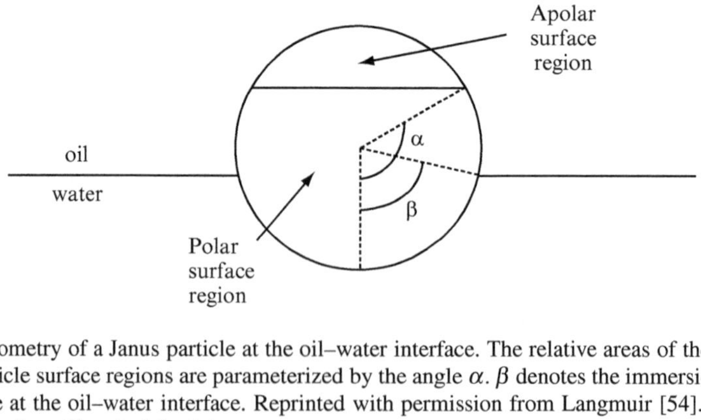

Fig. 9 Geometry of a Janus particle at the oil–water interface. The relative areas of the polar and apolar particle surface regions are parameterized by the angle α. β denotes the immersion angle of the particle at the oil–water interface. Reprinted with permission from Langmuir [54]. Copyright (2001) American Chemical Society

dumbbell-like CdS–FePt [61] and Ag–$CoFe_2O_3$ [62]. Recently, Duguet et al. showed the synthesis of an intermediate Janus particle composed of an inorganic part such as SiO_2 and an organic part consisting of PS, yielding dumbbell-like and snowman-like shapes [63]. Granick and coworkers successfully synthesized silica Janus colloid particles in large quantities at the liquid–liquid interface of molten wax and water [64, 65]. Moreover, Cohen et al. produced Janus-type microcapsules via LbL assembly followed by stamping an additional polymer layer onto one side of the LbL-capsule [62].

In addition to the efforts concerning the synthesis of Janus particles, their interfacial behavior was also of interest in order to verify Binks' predictions [54] experimentally. In this respect, it is a challenging problem – especially for polymer-based Janus structures – to clearly demonstrate the Janus character of the synthesized objects [66, 67]. The advantage of inorganic Janus particles is that they have a well-defined structure and geometry, which can often be visualized by electron microscopy. Thus, the interfacial properties of Janus particles can be directly compared to their homogeneous analogues. Recently, Glaser et al. [68] prepared Janus particles (Fig. 10) consisting of a gold part and an iron oxide part following a synthesis by Yu et al. [69]. For the Janus particles, the mean diameter of the gold particle is around 4 nm while the diameter of iron oxide is about 10 nm, resulting in an overall diameter of about 14 nm. (The diameters were determined by the image analysis program package ImageJ from the National Institutes of Health, Bethesda, MD). The homogeneous nanoparticles show a slightly smaller diameter compared to the overall size of the Janus type: the gold nanoparticle diameter is about 10 nm and the iron oxide nanoparticle diameter is about 7 nm (Fig. 10). Then, the particle amphiphilicity was tuned by ligand exchange with dodecanethiol or octadecanethiol on the gold part, and their interfacial activities were compared to those of the homogeneous gold and iron oxide particles using pendant drop tensiometry. The reduction in interfacial energy supported the theoretical predictions of Binks and Lumsdon (Fig. 11) [54, 68].

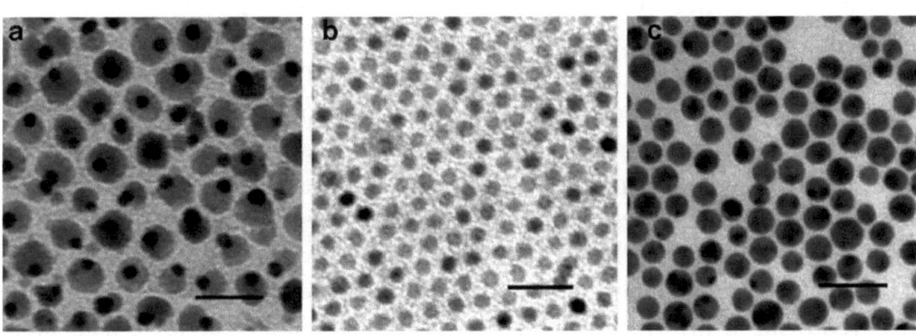

Fig. 10 TEM images of the nanoparticles: (**a**) Janus particles consisting of gold (*darker spheres*) and iron oxide (*brighter spheres*); (**b**) homogeneous iron oxide particles; (**c**) gold particles. Scale bars: 25 nm. Reprinted with permission from Langmuir [68]. Copyright (2006) American Chemical Society

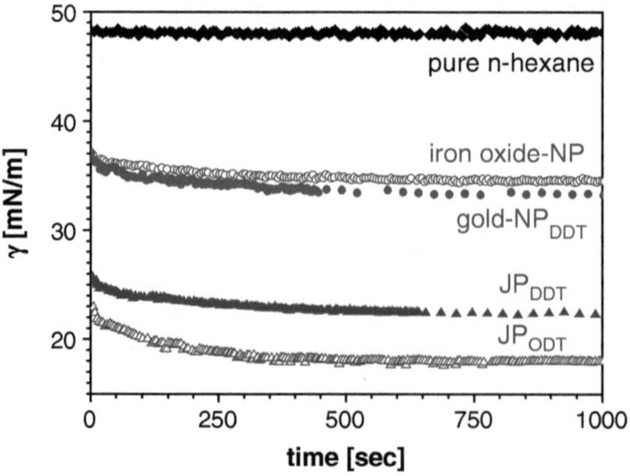

Fig. 11 Dynamic interfacial tension (γ) measurements of a hexane–water interface during adsorption of nanoparticles to a pendant water drop in hexane (for all particle types, concentration was 1.2×10^{-4} mmol/L). The gold moieties were modified using dodecanethiol (*DDT*) or octadecanethiol (*ODT*). *NP* homogeneous nanoparticles, *JP* Janus particles. Reprinted with permission from Langmuir [68]. Copyright (2006) American Chemical Society

1.5 Self-Assembly of Nanoparticle/Block Copolymer Mixtures

Incorporating nanoparticles into polymer matrices produces novel hybrid materials with special electrical, magnetic, and optical properties. One approach is to directly evaporate or synthesize inorganic nanoparticles inside well-ordered block polymer templates. Cohen [70], Sohn [71, 72], Möller [73] and coworkers have shown that polymer–nanoparticle composites can be generated by directly reducing the metal precursors inside block copolymer templates of different morphologies. Jaeger and coworkers demonstrated that by evaporating a series of metal nanoparticles on polystyrene-*block*-poly(methyl methacrylate) (PS-*b*-PMMA) templates, well-aligned metal wires could be obtained on the PS domains after thermal treatment at 180°C [74]. The disadvantages of these methods are lack of control of the size distributions of the nanoparticles, and lack of control of the position distributions of the nanoparticles inside the polymer matrices. Recently, Cheyne and Moffit demonstrated the formation of mesoscopic wires and cables via co-assembly of PS-decorated CdSe nanoparticles and a polystyrene-*block*-poly(ethylene oxide) (PS-*b*-PEO) block copolymer at the water–air interface [75]. Another approach was inspired by theoretical predictions of Balazs and coworkers [76–78], who calculated the morphology and thermodynamic behavior of copolymer–particle mixtures without requiring a priori knowledge of the equilibrium structures. The method combines a self-consistent field theory (SCFT) for the polymers with a density functional theory (DFT) for the particles. They applied this theory to diblock nanoparticle mixtures. The theory predicts ordered phases where particles and diblocks self-assemble into spatially periodic structures. They allowed the particles, by changing their solubility, to interact with both parts of the diblock copolymer

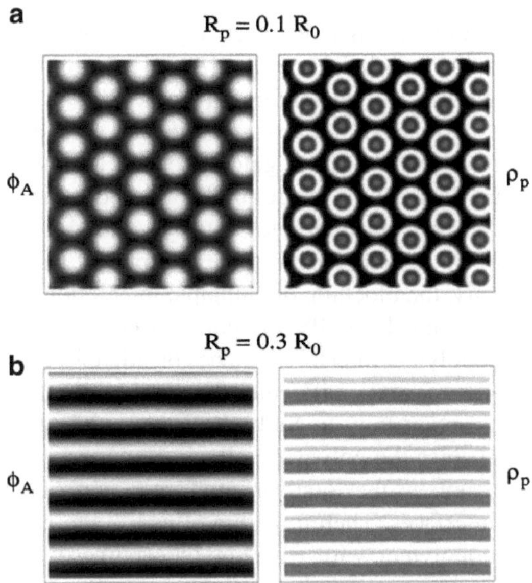

Fig. 12 Two-dimensional density plots for the diblock/particle systems obtained from the SCFT/DFT theory. Plots are for $\chi_{AB}N = 20$, $\chi_{AP} = \chi_{BP} = 0.02$, $\overline{N} = 1000$, $f = 0.35$, and $\phi_p = 0.15$. Plots on the *left* represent the distribution of the A blocks and plots on the *right* represent the distribution of the centers of mass of the particles. *Light regions* indicate a high density, while *dark regions* indicate low densities. The image in part (**a**) shows that the system displays a cylindrical phase when $R_p = 0.1R_0$ and the image in part (**b**) shows that the mixture forms a lamellar phase when the particle size is increased to $R_p = 0.3R_0$. Reprinted with permission from Macromolecules [77]. Copyright (2002) American Chemical Society

and found that variations in the particle–block interaction energies can induce phase transitions in the morphology of the mixture. At fixed interactions, variations in particle size can also induce similar transitions (Fig. 12).

Experimentally, Thomas and coworkers found that hydrocarbon-coated gold nanoparticles, with a diameter of 3.5 nm, segregated to the interface between the microdomains of poly(styrene-*block*-ethylene propylene) (PS-*b*-PEP) copolymer, whereas larger hydrocarbon-coated silica nanoparticles (21.5 nm in diameter) were located at the center of the PEP domains [79]. In the absence of specific enthalpic interactions between the two types of nanoparticles and the polymer matrix, the result suggests a profound influence of entropic contributions to the self-organization process. For large particles, the decrease in conformational entropy of the respective polymer subchains after particle sequestration is dominant, whereas for smaller particles, the decrease in entropy is outweighed by the particle translational entropy.

Manipulating the location of nanoparticles in the materials can also be achieved by controlling the surface properties of the nanoparticles. By using enthalpic interactions, Kramer and coworkers found that gold nanoparticles covered with sufficient thiol-terminated PS ligands are held inside the PS microdomains in polystyrene-*block*-poly(2-vinyl-piridine) (PS-*b*-P2VP) copolymers, whereas gold nanoparticles covered with both PS and P2VP ligands selectively segregate to the interfaces of PS and P2VP to minimize the interfacial energies [80, 81]. A similar phenomenon was also observed by Composto and coworkers for immiscible polymer blends in the presence of silica nanoparticles [82]. Subsequent reports demonstrated that controlling the ligand density on the surface of gold nanoparticles also greatly affects the nanoparticle positional distribution inside the PS microdomains. Strong enthalpic interaction between the gold nanoparticles and pyridine units in P2VP drive the nanoparticles with low PS ligand density to the interface of PS and P2VP, whereas at high PS ligand density, the blocking of enthalpic interaction results in a distribution of gold nanoparticles inside the PS microdomains [83, 84].

In the melt, block copolymer interfaces can be considered as "fluid", with dynamics much slower than for conventional liquids. However, the use of block copolymer interfaces provides an opportunity to study the nanoparticle assembly at fluid interfaces in more detail since the polymer melt can be quenched at any stage of the assembly process. This allows the "quasi in situ" study of nanoparticle assembly at the block copolymer interfaces. Therefore, thin films from mixtures of a cylindrical PS-*b*-P2VP diblock copolymer with tri-*n*-octylphosphine oxide-(TOPO)-covered CdSe nanoparticles were prepared and investigated with SFM, TEM, and GISAXS after thermal annealing. Surprisingly, these composite materials are found to form hierarchically ordered structures via a cooperative self-organization. On the one hand, the cylindrical microdomains of the copolymer dictate the spatial distribution of the nanoparticles within the film. On the other hand, nanoparticles are found to segregate to the interfaces, mediating interfacial interactions and surface energies, resulting in an orientation of the cylindrical domains normal to the surface, even when the interactions of one of the blocks with the substrate are strongly attractive (Fig. 13). Thus, the synergy between two assembly processes produces unique structures, without the use of external fields, and opens a novel route to new self-directing, self-assembling architectures [85].

To reveal the details of the cooperative self-organization of the PS-*b*-P2VP block copolymer/nanoparticle composites leading to the above-described hierarchically

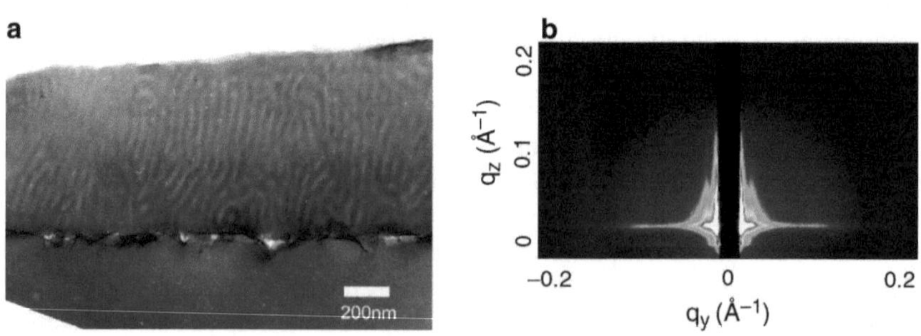

Fig. 13 (**a**) TEM image of cross-section of a PS-*b*-P2VP block copolymer/CdSe nanoparticle film after annealing at 170°C for 2 days. (**b**) Data from GISAXS measurements: of the same film at incident angle of 0.09° with a penetration depth of 61 Å. Reprinted with permission from Nature [85]. Copyright (2005) Nature Publishing Group

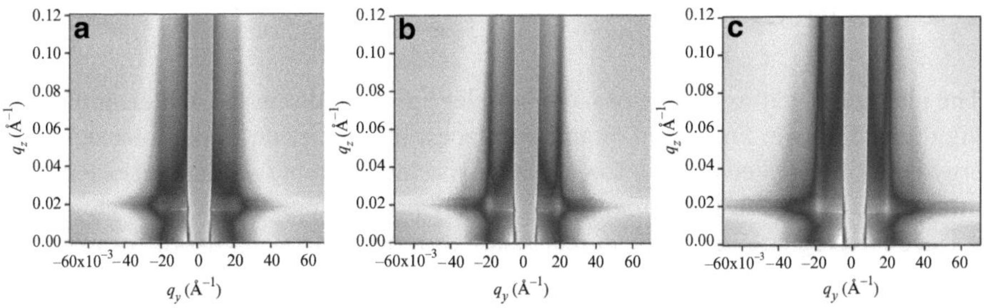

Fig. 14 Series of in-situ GISAXS images of a cylindrical nanoparticle-doped PS-*b*-P2VP film during thermal annealing at 180°C: (**a**) 0 h, (**b**) 4 h, and (**c**) 14 h after spin coating. Reprinted with permission from Advanced Materials [86]. Copyright (2007) Wiley-VCH

ordered structures, the structure formation process was investigated with in situ GISAXS during thermal annealing. It was found that the orientation of the microdomains starts at the free surface and propagates into the film (Fig. 14), while the CdSe nanoparticles segregate to the P2VP phase, filling the cylinders from the top. To demonstrate the universality of this process, lamellar microdomain morphologies were used in addition to cylindrical ones. The results are shown in the SEM image in Fig. 15, where the microdomains of a lamellar PS-*b*-P2VP block copolymer oriented perpendicular to the substrate can be seen. The inset depicts the decoration of the lamellae with CdSe nanoparticles (bright spots).

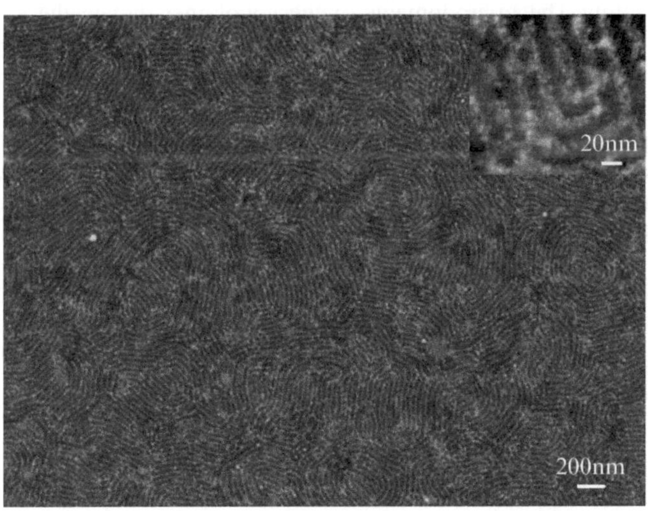

Fig. 15 High resolution SEM image of the lamellar PS-*b*-P2VP/CdSe nanoparticle composite thin film after annealing in $CHCl_3$ for 1 day, without staining. The image is taken at 1 kV acceleration voltage. Reprinted with permission from Advanced Materials [86]. Copyright (2007) Wiley-VCH

2 Conclusion

The interfacial self-assembly of nanoparticles has been discussed and examples are highlighted above. For liquid–liquid interfaces, the reduction in the interfacial energy is the dominated driving force. Nanoparticles assemble at the interface of two immiscible liquids in a disordered but densely packed monolayer. With increasing surface coverage, there exists an increasing difference between the diffusion at the early and late stages of adsorption. This reveals an energy barrier at late stages that corresponds to the activation energy for a thermally triggered escape of nanoparticles from the interface.

A promising application of the self-assembly of nanoparticles at droplet surfaces is the interfacial crosslinking of chemically functionalized nanoparticles. This enables the encapsulation of water-soluble or oil-soluble materials inside the resulting nanocontainers. By varying the concentration of reactive moieties, it will be possible to control the permeability and strength of these nanostructured membranes.

A special case of nanoparticle self-assembly is the Janus particle. It was shown that Janus particles are considerably more active than homogeneous particles of comparable size and chemical nature and that the interfacial activity can be increased by increasing the amphiphilic character of the particles. Thus, the Janus particles show a significant advantage in the stabilization of emulsions and foams over homogeneous particles as they unify the Pickering concept and the amphiphilicity of a simple surfactant.

For the self-assembly of nanoparticles in block copolymers, it was shown that these composite materials form hierarchically ordered structures via a cooperative self-organization: The microdomains of the copolymer dictate the spatial distribution of the nanoparticles within the film. The nanoparticles segregate to the interfaces, mediating interfacial interactions and surface energies, resulting in an orientation of the block copolymer domains normal to the surface. With respect to the wide range of synthetic methods for the production of well-defined nanoparticles of various types, the self-assembly of nanoparticles into hierarchically ordered structures, using interfacial interactions, represents a rich new area leading to potential applications in optical, acoustic, electronic, and magnetic materials.

Acknowledgments The authors thank C. Abetz, D.J. Adams, A. Balazs, D. Cookson, A. D. Dinsmore, T. Emrick, J. He, G. Krausch, X. Li, Y. Lin, S. Long, T.P. Russell, K. Sill, H. Skaff, R. Tangirala, J. Wang, Q. Wang, H. Xiang. This work was funded by the German Science Foundation (SFB 481, TP B10).

References

1. Binks BP (2007) Colloidal particles at liquid interfaces. Phys Chem Chem Phys 9(48):6298–6299
2. Pickering SU (1907) Emulsions. J Chem Soc 91:2001

3. Ramsden W (1903) Separation of solids in the surface-layers of solutions and 'suspensions' (observations on surface-membranes, bubbles, emulsions, and mechanical coagulation). Preliminary account. Proc R Soc Lond A Math Phys Eng Sci 72:156–164
4. Binks BP, Lumsdon SO (2000) Influence of particle wettability on the type and stability of surfactant-free emulsions. Langmuir 16(23):8622–8631
5. Pieranski P (1980) Two-dimensional interfacial colloidal crystals. Phys Rev Lett 45(7):569–572
6. Binks BP, Lumsdon SO (2001) Pickering emulsions stabilized by monodisperse latex particles: effects of particle size. Langmuir 17(15):4540–4547
7. Binks BP, Whitby CP (2004) Silica particle-stabilized emulsions of silicone oil and water: aspects of emulsification. Langmuir 20(4):1130–1137
8. Giermanska-Kahn J et al (2002) A new method to prepare monodisperse Pickering emulsions. Langmuir 18(7):2515–2518
9. Horozov TS et al (2003) Order-disorder transition in monolayers of modified monodisperse silica particles at the octane-water interface. Langmuir 19(7):2822–2829
10. Duan H et al (2005) Magnetic colloidosomes derived from nanoparticle interfacial self-assembly. Nano Lett 5(5):949–952
11. Duan HW et al (2004) Directing self-assembly of nanoparticles at water/oil interfaces. Angew Chem Int Ed 43(42):5639–5642
12. Lin Y et al (2003) Nanoparticle assembly and transport at liquid-liquid interfaces. Science 299(5604):226–229
13. Wang DY, Möhwald H (2004) Template-directed colloidal self-assembly – the route to 'top-down' nanochemical engineering. J Mater Chem 14(4):459–468
14. Schulman JH, Leja J (1954) Control of contact angles at the oil–water–solid interfaces – emulsions stabilized by solid particles (Baso4). Trans Faraday Soc 50(6):598–605
15. Aveyard R, Clint JH, Horozov TS (2003) Aspects of the stabilisation of emulsions by solid particles: effects of line tension and monolayer curvature energy. Phys Chem Chem Phys 5(11):2398–2409
16. Caruso F, Caruso RA, Möhwald H (1998) Nanoengineering of inorganic and hybrid hollow spheres by colloidal templating. Science 282(5391):1111–1114
17. Cha JN et al (2003) Spontaneous formation of nanoparticle vesicles from homopolymer polyelectrolytes. J Am Chem Soc 125:8285
18. Decher G (1997) Fuzzy nanoassemblies: toward layered polymeric multicomposites. Science 277(5330):1232–1237
19. Marinakos SM et al (1999) Gold particles as templates for the synthesis of hollow polymer capsules. Control of capsule dimensions and guest encapsulation. J Am Chem Soc 121(37):8518–8522
20. Tiarks F, Landfester K, Antonietti M (2001) Preparation of polymeric nanocapsules by miniemulsion polymerization. Langmuir 17(3):908–918
21. Caruso F et al (2000) Enzyme encapsulation in layer-by-layer engineered polymer multilayer capsules. Langmuir 16(4):1485–1488
22. Caruso F et al (2000) Microencapsulation of uncharged low molecular weight organic materials by polyelectrolyte multilayer self-assembly. Langmuir 16(23):8932–8936
23. Gao CY et al (2002) Spontaneous deposition of water-soluble substances into microcapsules: phenomenon, mechanism, and application. Angew Chem Int Ed 41(20):3789–3793
24. Fowler CE, Khushalani D, Mann S (2001) Interfacial synthesis of hollow microspheres of mesostructured silica. Chem Commun (19):2028–2029
25. Huo Q et al (1997) Preparation of hard mesoporous silica spheres. Chem Mater 9:14
26. Lu YF et al (1999) Aerosol-assisted self-assembly of mesostructured spherical nanoparticles. Nature 398(6724):223–226
27. Schacht S et al (1996) Oil-water interface templating of mesoporous macroscale structures. Science 273:768
28. Tricoli V, Sefcik J, McCormic AV (1997) Synthesis of oriented zeolite membranes at the interface between two fluid phases. Langmuir 13:4193

29. Dinsmore AD, Crocker JC, Yodh AG (1998) Self-assembly of colloidal crystals. Curr Opin Colloid Interface Sci 3(1):5–11
30. Dinsmore AD et al (2002) Colloidosomes: selectively permeable capsules composed of colloidal particles. Science 298:1006–1009
31. Mallwitz F, Goedel WA (2001) Physically cross-linked ultrathin elastomeric membranes. Angew Chem Int Ed 40(14):2645–2647
32. Xu H, Goedel WA (2002) Polymer-silica hybrid monolayers as precursors for ultrathin free-standing porous membranes. Langmuir 18(6):2363–2367
33. Xu H, Goedel WA (2003) From particle-assisted wetting to thin free-standing porous membranes. Angew Chem Int Ed 42:4694
34. Xu H, Goedel WA (2003) Particle-assisted wetting. Langmuir 19:4950
35. Chaikof EL (1999) Engineering and material considerations in islet cell transplantation. Annu Rev Biomed Eng 1:103
36. Joki T et al (2001) Continuous release of endostatin from microencapsulated engineered cells for tumor therapy. Nat Biotechnol 19:35
37. Lanza RP, Langer R, Vacanti J (2000) Principles of tissue engineering. Academic, San Diego
38. Read T-A et al (2001) Local endostatin treatment of gliomas administered by microencapsulated producer cells. Nat Biotechnol 19:29
39. Desai TA, Hansford DJ, Ferrari M (2000) Micromachined interfaces: new approaches in cell immunoisolation and biomolecular separation. Biomol Eng 17(1):23–36
40. Murray CB, Kagan CR, Bawendi MG (1995) Self-organization of Cdse nanocrystallites into 3-dimensional quantum-dot superlattices. Science 270(5240):1335–1338
41. Murray CB, Kagan CR, Bawendi (2000) Synthesis and characterization of monodisperse nanocrystals and close-packed nanocrystal assemblies. Annu Rev Mater Sci 30(1):545–610
42. Korgel BA (2004) Materials science – self-assembled nanocoils. Science 303(5662): 1308–1309
43. Shah PS et al (2003) Single-step self-organization of ordered macroporous nanocrystal thin films. Adv Mater 15(12):971–974
44. Sigman MB, Saunders AE, Korgel BA (2004) Metal nanocrystal superlattice nucleation and growth. Langmuir 20(3):978–983
45. Murray CB, Norris DJ, Bawendi MG (1993) Synthesis and characterization of nearly monodisperse Cde (E = S, Se, Te) semiconductor nanocrystallites. J Am Chem Soc 115(19):8706–8715
46. Lin Y et al (2005) Nanoparticle assembly at fluid interfaces: structure and dynamics. Langmuir 21(1):191–194
47. Lin Y et al (2003) Ultrathin cross-linked nanoparticle membranes. J Am Chem Soc 125(42):12690–12691
48. Kulak A et al (2003) Controlled assembly of nanoparticle-containing gold and silica microspheres and silica/gold nanocomposite spheroids with complex form. Chem Mater 15(2):528–535
49. Dai L, Sharma R, Wu CY (2005) Self-assembled structure of nanoparticles at a liquid-liquid interface. Langmuir 21:2641–2643
50. Kutuzov S et al (2007) On the kinetics of nanoparticle self-assembly at liquid/liquid interfaces. Phys Chem Chem Phys 9(48):6351–6358
51. Skaff H et al (2005) Crosslinked capsules of quantum dots by interfacial assembly and ligand crosslinking. Adv Mater 17(17):2082–2086
52. Tang ZY et al (2006) Self-assembly of CdTe nanocrystals into free-floating sheets. Science 314(5797):274–278
53. Jiang CY et al (2004) Freely suspended nanocomposite membranes as highly sensitive sensors. Nat Mater 3(10):721–728
54. Binks BP, Fletcher PDI (2001) Particles adsorbed at the oil-water interface: a theoretical comparison between spheres of uniform wettability and "Janus" particles. Langmuir 17(16):4708–4710
55. Perro A et al (2005) Design and synthesis of Janus micro- and nanoparticles. J Mater Chem 15(35–36):3745–3760

56. Erhardt R et al (2001) Janus micelles. Macromolecules 34(4):1069–1075
57. Erhardt R et al (2003) Amphiphilic Janus micelles with polystyrene and poly(methacrylic acid) hemispheres. J Am Chem Soc 125(11):3260–3267
58. Forster S, Antonietti M (1998) Amphiphilic block copolymers in structure-controlled nanomaterial hybrids. Adv Mater 10(3):195
59. Schrage S, Sigel R, Schlaad H (2003) Formation of amphiphilic polyion complex vesicles from mixtures of oppositely charged block ionomers. Macromolecules 36(5):1417–1420
60. Teranishi T et al (2004) Nanoacorns: anisotropically phase-segregated CoPd sulfide nanoparticles. J Am Chem Soc 126(32):9914–9915
61. Gu HW et al (2004) Facile one-pot synthesis of bifunctional heterodimers of nanoparticles: a conjugate of quantum dot and magnetic nanoparticles. J Am Chem Soc 126(18):5664–5665
62. Li ZF et al (2005) Layer-by-layer assembled Janus microcapsules. Macromolecules 38(19):7876–7879
63. Duguet E et al (2005) French Patent WO 2005/049195
64. Hong L, Jiang S, Granick S (2006) Simple method to produce Janus colloidal particles in large quantity. Langmuir 22(23):9495–9499
65. Jiang S et al (2008) Solvent-free synthesis of Janus colloidal particles. Langmuir 24(18): 10073–10077
66. Walther A, Matussek K, Muller AHE (2008) Engineering nanostructured polymer blends with controlled nanoparticle location using Janus particles. Acs Nano 2(6):1167–1178
67. Walther A, Muller AHE (2008) Janus particles. Soft Matter 4(4):663–668
68. Glaser N et al (2006) Janus particles at liquid-liquid interfaces. Langmuir 22:5227
69. Yu H et al (2005) Dumbbell-like bifunctional Au-Fe3O4 nanoparticles. Nano Lett 5(2):379–382
70. Boontongkong Y, Cohen RE (2002) Cavitated block copolymer micellar thin films: lateral arrays of open nanoreactors. Macromolecules 35(9):3647–3652
71. Sohn BH et al (2003) Directed self-assembly of two kinds of nanoparticles utilizing monolayer films of diblock copolymer micelles. J Am Chem Soc 125(21):6368–6369
72. Sohn BH, Seo BH (2001) Fabrication of the multilayered nanostructure of alternating polymers and gold nanoparticles with thin films of self-assembling diblock copolymers. Chem Mater 13(5):1752–1757
73. Spatz J et al (1998) Controlled mineralization and assembly of hydrolysis-based nanoparticles in organic solvents combining polymer micelles and microwave techniques. Adv Mater 10(6):473
74. Lopes WA, Jaeger HM (2001) Hierarchical self-assembly of metal nanostructures on diblock copolymer scaffolds. Nature 414(6865):735–738
75. Cheyne RB, Moffitt MG (2007) Controllable organization of quantum dots into mesoscale wires and cables via interfacial block copolymer self-assembly. Macromolecules 40(6): 2046–2057
76. Lee JY, Shou ZY, Balazs AC (2003) Predicting the morphologies of confined copolymer/nanoparticle mixtures. Macromolecules 36(20):7730–7739
77. Lee JY et al (2002) Effect of nanoscopic particles on the mesophase structure of diblock copolymers. Macromolecules 35(13):4855–4858
78. Thompson RB et al (2001) Predicting the mesophases of copolymer-nanoparticle composites. Science 292(5526):2469–2472
79. Bockstaller MR et al (2003) Size-selective organization of enthalpic compatibilized nanocrystals in ternary block copolymer/particle mixtures. J Am Chem Soc 125(18):5276–5277
80. Chiu JJ et al (2005) Control of nanoparticle location in block copolymers. J Am Chem Soc 127(14):5036–5037
81. Kim BJ et al (2005) Nanoparticle-induced phase transitions in diblock-copolymer films. Adv Mater 17(21):2618
82. Chung H et al (2005) Self-regulated structures in nanocomposites by directed nanoparticle assembly. Nano Lett 5(10):1878–1882
83. Kim BJ et al (2006) Effect of areal chain density on the location of polymer-modified gold nanoparticles in a block copolymer template. Macromolecules 39(12):4108–4114

84. Kim BJ et al (2007) Importance of end-group structure in controlling the interfacial activity of polymer-coated nanoparticles. Macromolecules 40(6):1796–1798
85. Lin Y et al (2005) Self-directed self-assembly of nanoparticle/copolymer mixtures. Nature 434(7029):55–59
86. He J et al (2007) Self-assembly of nanoparticle-copolymer mixtures: a kinetic point of view. Adv Mater 19:381–385

Adv Polym Sci (2010) 228: 59–121
DOI:10.1007/12_2009_35

Published online: 26 February 2010

Holographic Gratings and Data Storage in Azobenzene-Containing Block Copolymers and Molecular Glasses

Hubert Audorff, Klaus Kreger, Roland Walker, Dietrich Haarer, Lothar Kador, and Hans-Werner Schmidt

Dedicated to Prof. Dr. Helmut Ringsdorf on the occasion of his 80th birthday

Abstract This review covers synthesis, materials development, and photophysics of azobenzene-containing block copolymers as potential media for reversible volume holographic data storage. For high-density holographic data storage, volume gratings must be inscribed in millimeter-thick samples to achieve efficient angle multiplexing. It is demonstrated that block copolymers with azobenzene side-groups in the minority block develop no detrimental surface relief structures and exhibit superior performance regarding volume gratings, compared to homopolymers and statistical copolymers. Several material concepts for optimizing the refractive index modulation and the stability of volume gratings are presented. Stabilities of more than 2 years were achieved. Most important is the development of polymer blends comprising the azobenzene-containing block copolymer and an optically transparent homopolymer. This enables the preparation of millimeter-thick samples with the required optical density of ~0.7 at the writing wavelength by conventional injection molding techniques. The inscription of up to 200 holograms at the same lateral position was demonstrated. In addition, more than 1,000 write/erase cycles can be performed. This is the first time that the inscription and erasure of the long-term

H. Audorff, D. Haarer, and L. Kador (✉)
Institute of Physics, Bayreuther Institut für Makromolekülforschung (BIMF)
University of Bayreuth, 95440 Bayreuth, Germany
e-mail: lothar.kador@uni-bayreuth.de

K. Kreger, R. Walker, and H.-W. Schmidt (✉)
Macromolecular Chemistry I, Bayreuther Institut für Makromolekülforschung (BIMF)
University of Bayreuth, 95440 Bayreuth, Germany
e-mail: hans-werner.schmidt@uni-bayreuth.de

stable angle-multiplexed volume gratings in a rewritable polymeric medium have been achieved by purely optical means.

A second important application for azobenzene-containing materials is the controlled preparation of surface relief structures. It is demonstrated that azobenzene-containing molecular glasses are an ideal class for efficient formation of surface relief gratings (SRGs) with amplitude heights of more than 600 nm. Clear relationships can be established between the chemical structure of the molecules and the behavior of SRG formation. All results are in agreement with the gradient force model by Kumar et al. The surface patterns are stable enough to be transferred to a polymer surface via replica molding.

Keywords Angle multiplexing · Azobenzene · Diblock copolymers · Holographic data storage · Molecular glasses · Phase gratings · Surface relief gratings

Contents

Abbreviations and Symbols

45°	Electric field vector at an angle of 45° to the plane of incidence
AFM	Atomic force microscopy
AIBN	Azobisisobutyronitrile
ATRP	Atom transfer radical polymerization
BS	Beam splitter
c	Cylindrical
CCD	Charge-coupled device
CD	Compact disc

d_0	Thickness of the film
Δd_{max}	Maximum SRG height
DSC	Differential scanning calorimetry
$\vec{E}$	Electric field vector of the incident light
f	Gradient force
GPC	Gel permeation chromatography
h	Planck constant
HEMA	Hydroxyethylmethacrylate
HOE	Holographic optical element
I	Intensity
J_1	First order Bessel function
k	Absorption coefficient
k_b	Boltzmann constant
λ	Wavelength of laser beam
lcp	Left circularly polarized light
m	Miscible
M_n	Number average molecular weight
M_w	Weight average molecular weight
η	Diffraction efficiency
n	Refractive index of the sample
n_1	Refractive index modulation
$n_{1,max}$	Maximum refractive index modulation
Δn	Amplitude of the refractive index change between sample and air
NEXAFS	Near-edge X-ray absorption fine structure
NMP	Nitroxide mediated polymerization
NMR	Nuclear magnetic resonance
ν	Frequency
OD	Optical density
p	Electric field vector parallel to the plane of incidence
P	Polarizer
P	Polarization induced by the electric light field
PAP	Photoaddressable polymers
PB	Polybutadiene
PDI	Polydispersity index
PDMS	Poly(dimethyl)siloxane
PMMA	Polymethylmethacrylate
POLMIC	Polarized light microscopy
PS	Polystyrene
rcp	Right circularly polarized light
ROM	Read-only memory
ru	Repeating unit
s	Sphere
s	Electric field vector perpendicular to the plane of incidence
S′	Sensitivity

SAXS	Small angle X-ray scattering
SRG	Surface relief grating
τ_1	Time constant of the build-up of the volume grating
τ_2	Time constant of the build-up of the surface relief grating
t	Writing time
T	Temperature
T_g	Glass transition temperature
TEM	Transmission electron microscopy
TGA	Thermogravimetric analysis
THF	Tetrahydrofuran
UV/vis	Ultraviolet/visible
w_A	Normalized weight of block A
WAXS	Wide-angle X-ray scattering
WORM	Write-once read-many
θ	Angle of incidence of the laser beam
$\Delta\varphi$	Phase difference between volume and surface relief grating
χ	Electrical susceptibility

1 Introduction

In the past few decades, the amount of information to be stored has rapidly increased, and will increase even more dramatically in the future. This development has led to an ongoing effort in searching for new concepts and materials for data storage. Optical data storage is commercially well established and the storage capacity and data transfer rate have substantially increased from CD-ROM to multilayer Blu-ray discs. With the future demand for storage capacities and transfer rates, the concept of two-dimensional optical storage employing only stacked layers will eventually encounter its limits and the step to new three-dimensional data storage is required.

The formation of holographic gratings, especially volume gratings, is one of the most promising techniques in this field. This method makes use of the entire volume instead of a single layer, so that storage capacities in the terabyte range on the size of a conventional compact disc will be achievable. Write-once media for holographic data storage are mainly based on photopolymers [1–6]. Generally, this concept relies on a diffusion-driven process of a monomeric species caused by light-induced initiation of a polymerization in defined regions. The storage mechanism – a refractive index modulation – arises from the created concentration gradient between exposed, photo-polymerized areas and the depleted surroundings. In contrast to this class, photochromic materials are rewritable. The most important class are azobenzene compounds. Owing to the rich photochemistry of azobenzene chromophores, a large variety of small-molecular and polymeric compounds has been synthesized for use as smart light-responsive materials for various potential applications, which is summarized in a recently published book by Zhao and Ikeda [7]. Most widely known is the effect that, upon light absorption, the azobenzene moieties

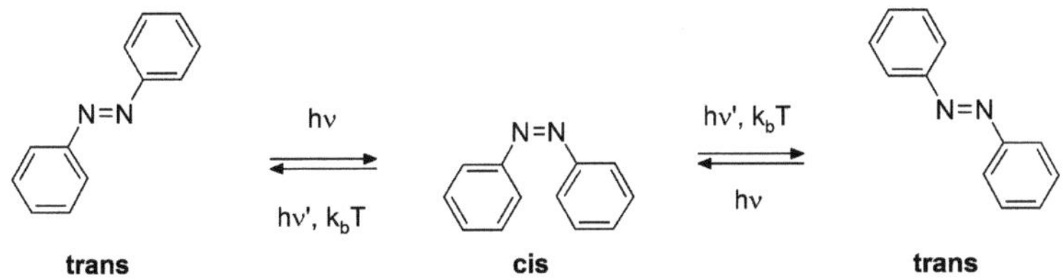

Fig. 1 Schematic representation of the reversible *trans–cis–trans* photo-isomerization of azobenzene. Upon absorption of UV light, a *trans*-to-*cis* conversion is induced; with visible (blue) light, a *cis*-to-*trans* conversion occurs until an equilibrium between both isomers is reached. The thermodynamically less stable *cis* isomer can thermally return to the *trans* isomer

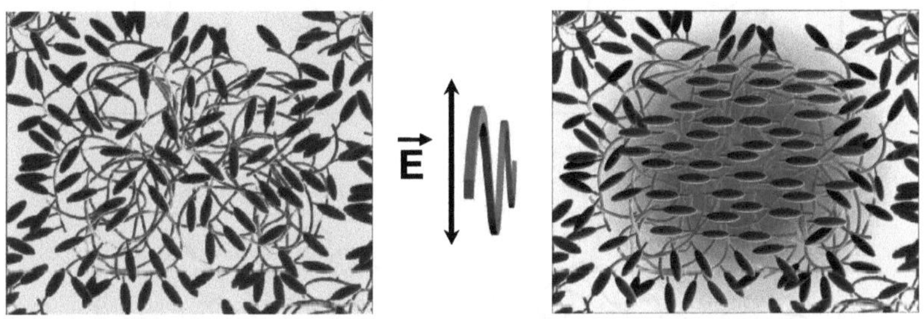

Fig. 2 Schematic representation of the reorientation of azobenzene chromophores by excitation with linearly polarized light. Since both the *cis* and the *trans* state are excited by illumination with the writing wavelength, the molecules perform repeated reorientations until they end up with their transition dipole moments perpendicular to the light polarization. (Adapted with permission from [8]. Copyright Wiley-VCH Verlag GmbH & Co. KGaA)

can reorient in the solid state (see Fig. 1) through multiple reversible *trans–cis–trans* photo-isomerization cycles.

With polarized light this process continues until the molecular transition dipole moment of the azobenzene is turned perpendicular to the exciting laser field and electronic excitations are no longer possible. As a consequence, the exposed areas become macroscopically birefringent. Since the shape-anisotropic azobenzene moiety has different polarizabilities parallel and perpendicular to its axis, the exposed areas have a different refractive index compared to the non-exposed areas (see Fig. 2). Illumination with a light intensity pattern caused by two interfering coherent laser beams thus creates a holographic phase grating.

Tailoring a suitable material for applications in holographic data storage is a challenging task. A good holographic storage material should fulfill a combination of required properties including, for example, high photosensitivity, the possibility of preparing thick samples with a sufficiently low optical density, high optical quality, high dynamic range, high-fidelity recording, and sufficient long-term stability.

Azobenzene-containing materials can be classified into two groups: polymer-based systems and molecular glasses. Azobenzene chromophore units, which are the optically addressable moiety, can be incorporated into polymer systems in several ways, as is schematically illustrated in Fig. 3.

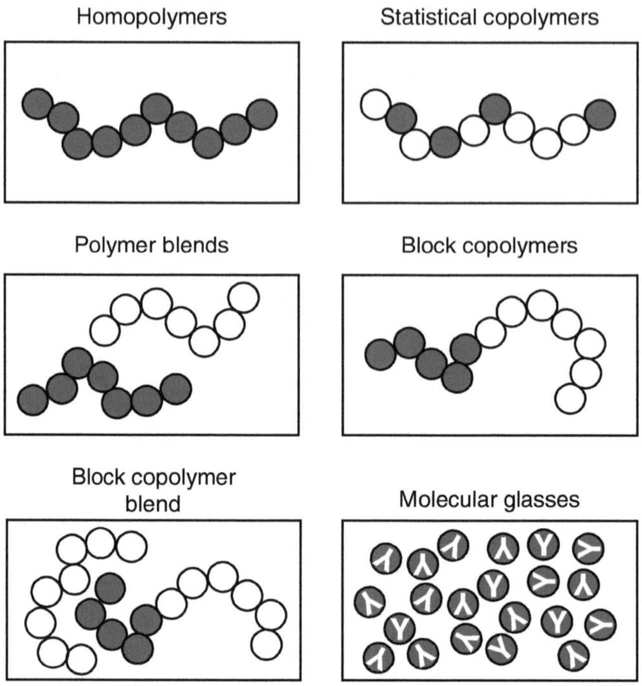

Fig. 3 Schematic representation of different polymer materials and molecular glasses containing azobenzene moieties. Azobenzene chromophore units *(shaded circles)*; non-absorbing repeating units of the polymer *(open circles)*; azobenzene-containing molecular glasses (*shaded circles enclosing* "Y")

With respect to the application as three-dimensional holographic storage material or as material for surface structuring, these materials behave quite differently. For holographic data storage utilizing the volume, thick samples are required in order to perform angle multiplexing with a high angular selectivity. To this end the optical density has to be sufficiently low, so that the laser beam can penetrate the sample. This is not achievable with homopolymers and molecular glasses, since they exhibit an optical density far too high. On the other hand, in order to utilize efficiently the *trans–cis–trans* isomerization for enhanced writing speeds and improved stabilization, the azobenzene chromophores have to be close to each other to benefit from the so-called "cooperative effect". Several ways to decrease the optical density were investigated. Diluting the chromophores by statistical copolymerization is not a real option since the cooperative effect and, hence, the long-term stability of the grating are lost. Therefore, block copolymers with microphase separation are the most promising class for holographic storage.

The formation of surface relief gratings (SRGs) is detrimental to volume gratings, if three-dimensional data storage is desired. Since SRGs are thin gratings, they have no angular selectivity. Moreover, their diffraction efficiency is usually higher than those of phase gratings in the volume. The formation of SRGs is most pronounced on the surfaces of homopolymers and glasses. Copolymers also show this effect, albeit to a smaller degree.

Homopolymers with azobenzene units as side-groups were first introduced by Ringsdorf and Schmidt et al. [9, 10]. Here, the azobenzene units were attached to the polymer backbone as side-groups via a flexible spacer. These polymers form thermotropic liquid crystalline phases. Their applicability as a holographic data storage medium was described by Eich et al. and later by various other groups [11–13]. Over the years, many different polymer systems were synthesized and the *trans–cis–trans* photo-isomerization of homopolymers was also investigated with respect to applications as optical data storage materials. For instance, Hvilsted et al. reported the synthesis and characterization of various polyesters with liquid crystalline azobenzene side-groups [14] and also of azobenzene-containing peptide oligomers [15]. Statistical copolymers with azobenzene dyes attached to the polymer backbone were first investigated as dye-containing liquid crystalline polymers [9, 16]. About 10 years later they were investigated in view of holographic data storage. Natansohn et al. investigated different types of azobenzene-containing statistical copolymers. Statistical methacrylate copolymers with azobenzene side-groups were also studied by Zilker et al. [17, 18] and by Hagen et al. [8] However, both homopolymers and statistical copolymers tend to develop SRGs in addition to the volume phase gratings during holographic exposure [19, 20]. Both properties in combination with the too high optical density exclude them from volume holography data storage applications.

Almost all polymer blends consisting of two or more polymers form macrophase-separated morphologies with sizes in the micrometer range. Thus they show bulk light scattering and are therefore not suitable for holographic data storage. One literature example is known, however, where this issue was overcome and multiplexing of 20 holograms could be accomplished. In this case the blend system was realized from two polymers with similar backbones [21].

Block copolymers composed of an amorphous block and a block containing the azobenzene side-groups are of interest because they form microphase-separated, uniform, and regular morphologies below 100 nm. Hence, in contrast to polymer blends, they do not scatter visible light. In addition, the cooperative effect is maintained in the confined geometry. Consequently, the interest in block copolymers with azobenzene chromophores has increased substantially over the past years. Our work on azobenzene-containing block copolymers and blends of azobenzene-containing block copolymers with amorphous non-absorbing homopolymers with respect to holographic volume storage will be reviewed in this chapter. Several other research groups have also reported on block copolymers carrying azobenzene chromophores in one block. They were mostly synthesized by controlled radical polymerization techniques such as atom transfer radical polymerization (ATRP) [22–41] or nitroxide mediated polymerization (NMP) [42]. Alternative approaches like a subsequent polymeranalogous reaction on the anionic preformed polymerbackbone have also been reported [43–45]. Only very few holographic experiments, especially with respect to angle multiplexing, have been reported, however.

An interesting new class and an alternative to polymers are small molecules forming stable amorphous phases, so-called molecular glasses. They represent an emerging material class of great potential for electronic, optical, and electro-optical applications. Such materials are already used in a number of devices, e.g., in photoconductor drums, organic light emitting devices, organic solar cells, photorefractive materials, and anti-reflective coatings [46–53]. In contrast to functional polymers, molecular glasses have the advantages of a well-defined molecular structure, the absence of structural defects, the absence of undefined end groups, and a uniform molecular weight. Molecular glasses with glass transition temperatures (T_g) well above room temperature are obtainable. Non-planar and space-filling molecular structures are common features used to suppress crystallization.

Molecular glasses can be envisioned for the same applications as photoaddressable polymers, but they have been studied far less extensively. It has been demonstrated that photoinduced birefringence can also be generated in thin films of azobenzene-based molecular glasses [54–58]. Like homopolymers, they form SRGs in addition to gratings in the volume when illuminated with two interfering laser beams.

In addition to the above-mentioned polymer-based blends, azobenzene-containing sol–gel-based systems are a third material class. It was shown by Stumpe et al. that an efficient SRG formation can be obtained using ionic azobenzene units, which were stabilized in a polyelectrolyte matrix via an in situ sol–gel reaction [59].

This review on our research on azobenzene-containing block copolymers and molecular glasses for holographic gratings and data storage is structured as follows. Section 2 outlines general synthetic pathways to azobenzene-containing block copolymers and molecular glasses, gives examples, and reports on the basic physical properties. In Sect. 3, the optical setup with its variations for the generation and characterization of holographic gratings is briefly described. Section 4 is devoted to SRG formation in azobenzene-containing materials and covers structure-property relations with respect to an efficient and controlled formation of SRGs as well as methods to suppress SRGs. Section 5 covers holographic volume gratings based on azobenzene-containing block copolymers and structure-property relations with respect to an improvement of the refractive index modulation and the long-term stability. Section 6 describes features of volume holograms for data storage applications in more detail. The angle multiplexing of thick gratings and the parallel inscription of entire pages of bit patterns with a spatial light modulator (SLM) are addressed, which are prerequisites for high-density data storage.

2 Materials: Design, Synthesis, and Properties

2.1 Azobenzene-Containing Block Copolymers

Block copolymers with liquid crystalline side chains were investigated for the first time by Adams and Gronski in 1989 [60]. A polymeranalogous reaction was used to attach the mesogenic side-groups to one of the blocks. In a similar way, block copolymers with azobenzene chromophores were synthesized by Ober et al. [43, 44]. In the scope of this review the synthesis of azobenzene-containing block

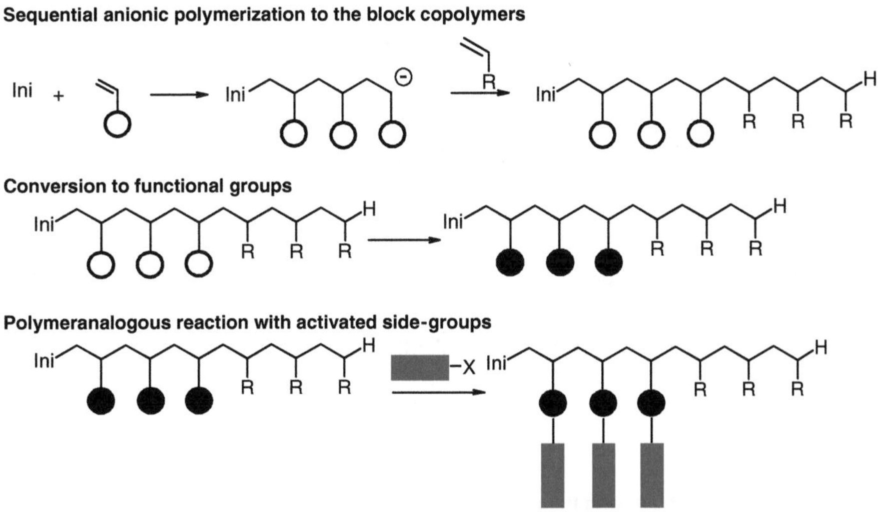

Fig. 4 General synthetic strategy to obtain block copolymers with azobenzene-containing side-groups via polymeranalogous reaction; *(open circles)* protected functional groups, *(filled circles)* functional groups, *(rectangles)* azobenzene chromophore units

copolymers, which combine the optical properties of azobenzene chromophores with the properties of block copolymers, will be discussed. Such block copolymers are hardly accessible by the direct living polymerization technique of corresponding azobenzene monomers. Therefore the synthetic strategy via a polymeranalogous reaction with one of the blocks was utilized. This approach is schematically illustrated in Fig. 4. It includes several advantages. For instance the monomers used for the anionic polymerization can be easily purified according to well-known procedures. This is in contrast to monomers, which already bear the azobenzene chromophore moiety and other polar substituents. Such monomers cannot be easily purified and polymerized under living anionic conditions due to possible termination, side, and transfer reactions during polymerization. In addition, block copolymers based on commercial monomers can be easily prepared in laboratory quantities in the range of 100–200 g, allowing several polymeranalogous reactions with azobenzene side-groups to be performed.

It is of great advantage to activate the low-molecular-weight side-groups instead of the polymer. Activation of the polymer or the block copolymers often results in crosslinked and insoluble materials. Low-molecular-weight azobenzene-containing side-groups can be transformed into an activated species via common organic chemistry reactions. Furthermore, the polymeranalogous reaction allows simple variation of the side-groups on a given block copolymer as well as easy access to statistically substituted blocks, if two chemically different side-groups are chosen.

In Fig. 5 the synthesis of a block copolymer system based on polystyrene (PS) and hydroborated 1,2-polybutadiene (PB) is shown. Polystyrene and 1,2-polybutadiene were chosen due to the well established polymerization conditions and techniques of the two monomers. Polystyrene-*block*-1,2-polybutadiene

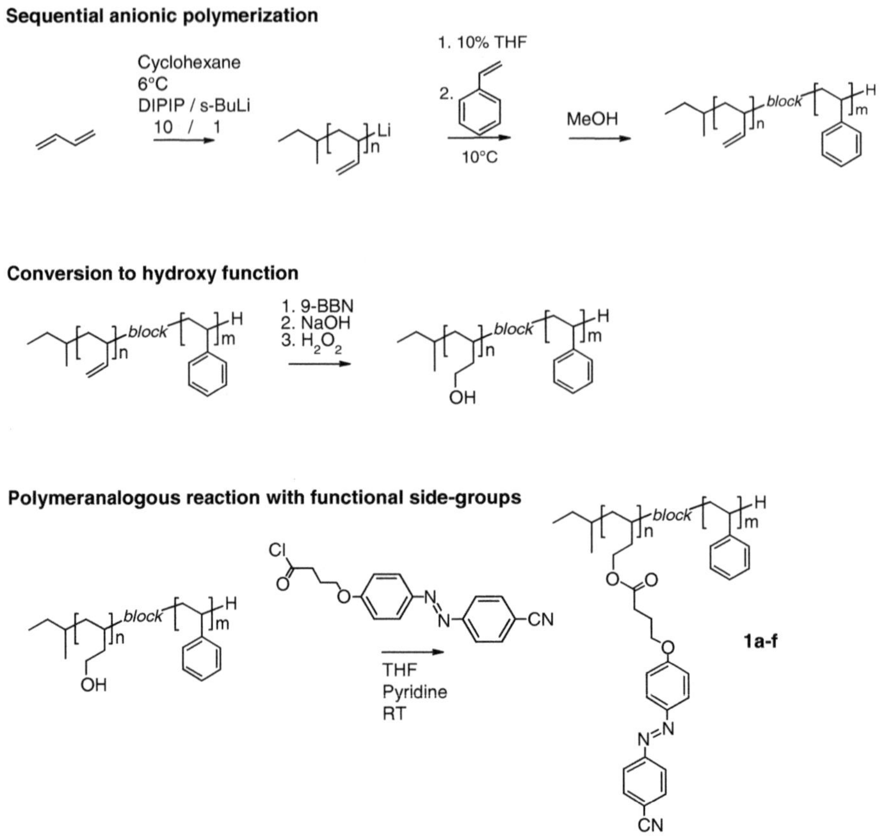

Fig. 5 Schematic representation of the block copolymer synthesis using the example of poly(2-(2-(4-(4-(4-cyano-phenylazo)-phenoxy)-butyric acid)ethyl ester)ethylene)-*block*-polystyrene. (Adapted with permission from [61]. Copyright Wiley-VCH Verlag GmbH & Co. KGaA)

block copolymers were synthesized according to a procedure described by Bates et al. [62] and Sänger et al. [63]. The polymerization was started with the stereospecific synthesis of 1,3-butadiene in cyclohexane utilizing dipiperidinoethane as complexation agent to yield more than 95% of the 1,2-polybutadiene. Tetrahydrofuran (THF) was added to break the complexation of the living chain ends and styrene was added to prepare the second block. The molecular weights were determined by gel permeation chromatography utilizing THF as eluent and narrowly distributed polystyrene standards.

Subsequently the polybutadiene segment was functionalized by quantitatively converting the olefin double bonds of the 1,2-polybutadiene block via a hydroboration reaction to hydroxyl functions. In the last step various azobenzene-chromophores were attached via a polymeranalogous reaction. The synthesis and characterization as well as photophysical aspects of block copolymers based on an amorphous polystyrene block and a functionalized 1,2-polybutadiene block bearing methoxyphenylazobenzene chromophores have been described in detail previously [61]. Homopolymers as well as block copolymers bearing methoxy-azobenzene side-groups are not liquid crystalline. We present here the synthesis and characteristic data of block copolymers with covalently linked cyano-azobenzene chromophores as side-groups, which are liquid crystalline and form a nematic mesophase [64].

The characteristic polymer data for a series of block copolymers with cyano-azobenzene are summarized in Table 1. Within the presented block copolymer series **1** the molecular weight of **1a–e** is similar with approx. 500–600 styrene repeating units. Block copolymer **1f** has approximately twice the molecular weight. The azobenzene content of the minority blocks varies from 27.4 to 1.9 wt%. The thermal behavior was investigated by differential scanning calorimetry (DSC), thermogravimetric analysis (TGA) and polarized light microscopy (POLMIC). The DSC curves of the block copolymers **1a–d** are shown in Fig. 6. For **1a–c** two glass transitions were observed, whereas for **1d** only the glass transition of polystyrene could be detected. The T_gs at ~52°C are assigned to the azobenzene minority blocks, which are approximately at the same temperature as determined for the corresponding homopolymer (T_g: 49°C). The cyanoazobenzene-based homopolymer was prepared in a similar way as described above starting with the anionic polymerization of 1,2-polybutadiene, conversion to hydroxyl-functions and polymeranalogous reaction with the azobenzene moieties (M_n: 5,100 g mol^{-1} corresponding to approximately 17 repeating units, PDI: 1.15, >98% conversion). The second T_g, located at ~102°C, is observed for all block copolymers and attributed to the polystyrene majority block. Block copolymer **1a** exhibits additionally a peak at 87°C (homopolymer: 89°C) which is assigned to the melting point of the semi-crystalline side chains. The melting peak of **1a** is the transition temperature into a liquid crystalline phase. The transition to the isotropic phase was found at a temperature of 180°C. The homopolymer exhibits a lower clearing temperature at 137°C. This increase of the clearing temperature is attributed to the higher degree of polymerization of the block containing the azobenzene chromophors in the block copolymer compared to the homopolymer. Similar findings were demonstrated by Stevens et al. [65]. For **1b** a clearing temperature of 150°C was detected, whereas the other block copolymers do not reveal a liquid crystalline phase in DSC. Annealing experiments of **1a** at 172°C revealed in the POLMIC that a fine birefringent texture is formed, which confirms the liquid crystalline phase behavior.

For comparison, a statistical copolymer was synthesized by a common free radical polymerization of hydroxyethyl methacrylate (HEMA) and styrene with AIBN as initiator, followed by a polymeranalogous reaction to the azobenzene-containing polymer (M_n: 33,000 g mol^{-1}, PDI: 1.6). The copolymer, comprising 28 wt% of azobenzene side-groups, exhibits a T_g at 88°C, but no liquid crystalline behavior was observed.

In order to prove the microphase separation of both blocks, which is required to maintain the cooperative effect of the azobenzene chromophores, thin films of **1a** (cyanoazobenzene content: 27.4 wt%) were investigated by transmission electron microscopy (TEM; see Fig. 7) and small angle X-ray scattering (SAXS). The

Table 1 Characteristic properties of the block copolymer series based on a polystyrene block and a 1,2-polybutadiene block containing cyanoazobenzene side-groups

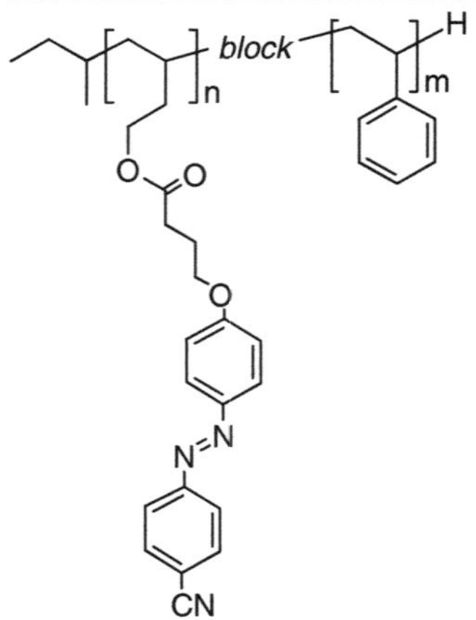

	w_{PS}[a]	w_R[a]	ru_{PS}[b]	ru_R[b]	M_n	M_w/M_n	T_g(°C)	Morph.
1a	72.6	27.4	533	61	77,900	1.15	51,102	*c*
1b	82.5	17.5	540	33	68,000	1.15	56,102	*c*
1c	90.0	10.0	528	17	61,000	1.13	58,103	*s*
1d	94.1	5.9	606	11	67,000	1.11	102	*s*[c]
1e	97.6	2.4	506	4	54,000	1.04	100	*m*
1f	98.1	1.9	1,102	7	117,000	1.12	104	*m*

[a] Weight fraction (%): average values obtained by NMR and UV-vis measurements
[b] Repeating units: average values from NMR and UV–vis on the basis of GPC measurements
[c] Morphology is not well-defined; *c* cylindrical, *s* spheres, *m* miscible

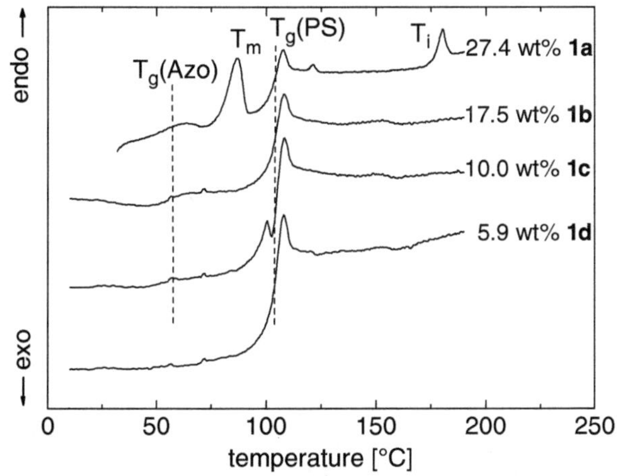

Fig. 6 DSC curves of cyanoazobenzene-containing block copolymers **1a–d** with different weight fraction of the azobenzene block. Second heating curves at a rate of $10\,K\,min^{-1}$ are shown

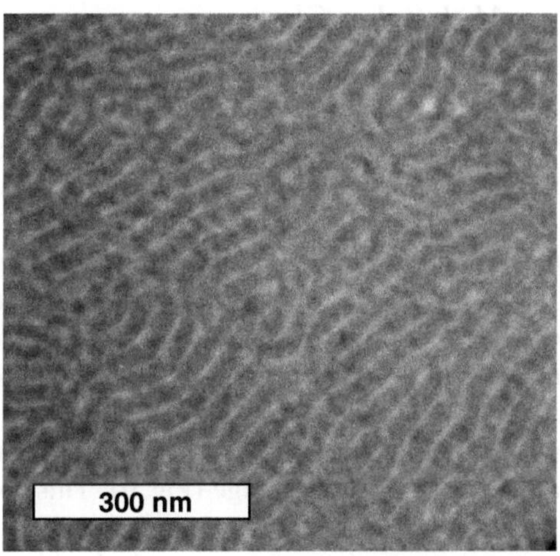

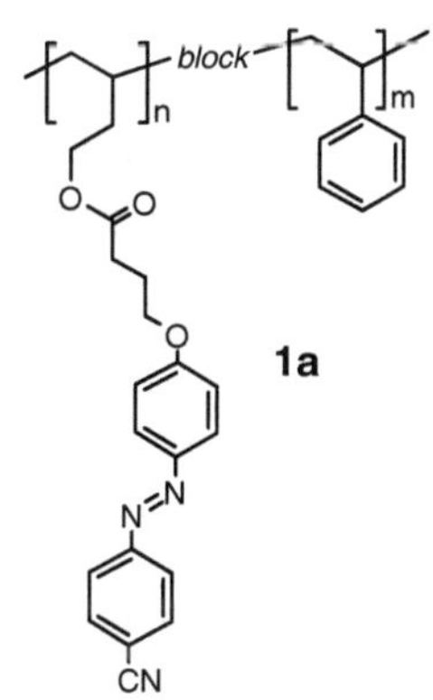

1a

Fig. 7 Electron transmission micrograph for **1a** (M_n: $76{,}900\,g\,mol^{-1}$; PDI: 1.15; minority segment: 27.4 wt%) cast from toluene, annealed three days at 120°C, and stained with RuO_4 (60 min)

polystyrene matrix was preferentially stained with RuO_4 vapor and, therefore, appear black in the pictures. This is in agreement with the observations of Mao et al. [43]. The cyanoazobenzene domains were transparent to the electron beam and, therefore, appear as the bright minority phase. A cylindrical morphology was identified with a domain spacing (distance between cylinders) of 43 nm which is in very good agreement with the lattice distance calculated of the SAXS profile (42 nm). The cylinder diameter was detected to be 13 nm. A cylindrical morphology was observed for block copolymer **1b**. Block copolymer **1c** was assigned to a spherical morphology as expected for a 9.3 wt% minority block content. The TEM picture of the stained film sample of block copolymer **1d** had very low contrast; therefore, the assignment of the morphology was difficult. Block copolymer **1d** was also assigned to a spherical morphology. In the X-ray diffraction pattern only one peak is observed at $q = 0.26\,\text{nm}^{-1}$ for **1c** corresponding to a lattice distance of 24 nm. Block copolymer **1d** was similar ($q = 0.30\,\text{nm}^{-1}$; lattice distance: 21 nm). For the block copolymers **1e** and **1f** no indications of a phase separation were observed, neither in the TEM micrograph nor in the SAXS measurements. It should be noted that both block copolymers behave during the optical and holographic measurements similar to statistical copolymers, supporting the existence of a non-phase-separated solid state.

Azobenzene-functionalized AB block copolymer systems are not limited to polystyrene-poly-1,2-butadiene-based block copolymers. In addition, we have prepared similar AB block copolymers, which are based on an amorphous PMMA matrix block and an azobenzene-functionalized PHEMA block [66] and on a PS matrix block and a functionalized PHEMA block [67].

2.2 Azobenzene-Containing Molecular Glasses

Azobenzene-containing molecular glasses were first introduced by Shirota et al. in 1998 [68]. One azobenzene chromophore was linked to an arylamine derivative. Since then some azobenzene-containing molecular glasses with different topologies and a different number of chromophores have been reported [55, 57, 69, 70]. In the following we present the synthesis and the characterization of molecular glasses with azobenzene chromophores which are based on different topologies with two, three, or four azobenzene moieties as illustrated in Fig. 8.

Most of the azobenzene-containing molecular glasses known from literature are synthesized by aryl–aryl and N–aryl coupling reaction. In contrast to this, the synthesis of our azobenzene-based molecular glasses is accomplished by linking azobenzene chromophores to the cores by an ester linkage (Fig. 9). This route includes several advantages. The formation of the ester linkages between the core and the side-groups allows easy access to a large variety of compounds. In this case the core (**2** and **4**) can carry the hydroxyl functions or the acid functions (**3**). In addition, commercially available compounds can be used, e.g., 4-hydroxyazobenzene and

Fig. 8 Schematic structure of different topologies of photoaddressable azobenzene containing molecular glasses with two, three, and four azobenzene moieties; *(open circles)* core; *(filled rectangles)* azobenzene chromophore units

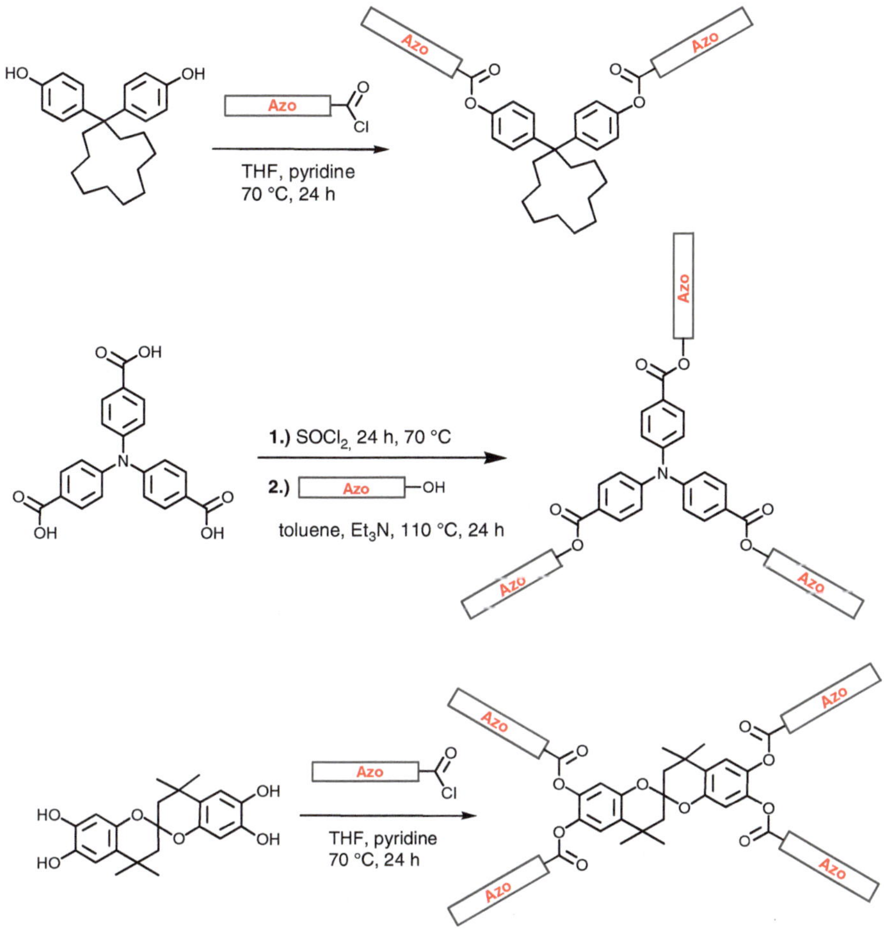

Fig. 9 Synthetic route to the photoresponsive azobenzene containing molecular glasses with two, three, and four azobenzene moieties

the acid chloride of 4-carboxyazobenzene. More complex substituted azobenzene moieties are easily accessible by utilizing common azobenzene coupling reactions. Using these substituted azobenzenes, it is possible to tailor the thermal and optical properties of the azobenzene-containing molecular glasses.

Typical examples of azobenzene-containing molecular glasses corresponding to the topologies given in Fig. 8 are shown in Fig. 10. These compounds are based on different core compounds with two, three, or four arms of unsubstituted azobenzene chromophores. The thermal and morphological properties of the compounds were investigated by DSC, TGA, POLMIC, and X-ray diffraction. The characteristic thermal properties of **2–4** are given in Table 2 and representative DSC curves are shown in Fig. 11.

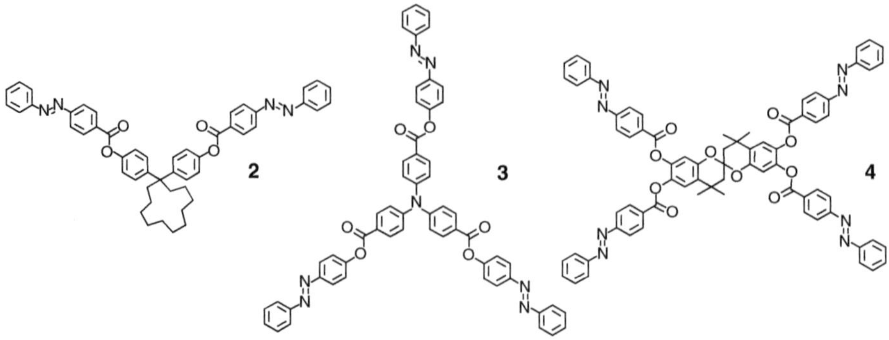

Fig. 10 Structures of the azobenzene-based molecular glasses **2**–**4** with two, three, and four azobenzene moieties

Table 2 Thermal properties of the photochromic molecular glasses

Compound	MW ($g\,mol^{-1}$)	First heating[a] T_m (°C)	First heating[a] T_{cryst} (°C)	Second heating[a] T_g (°C)	Second heating[a] $T_{recryst}$ (°C)	Second heating[a] T_m (°C)
2	757	234	n.d.	98	n.d.	n.d.
3	918	160,175	n.d.	92	n.d.	n.d.
4	1,205	257	n.d.	109	212	257

[a]Obtained from DSC: heating and cooling rate: $10\,K\,min^{-1}$ under N_2

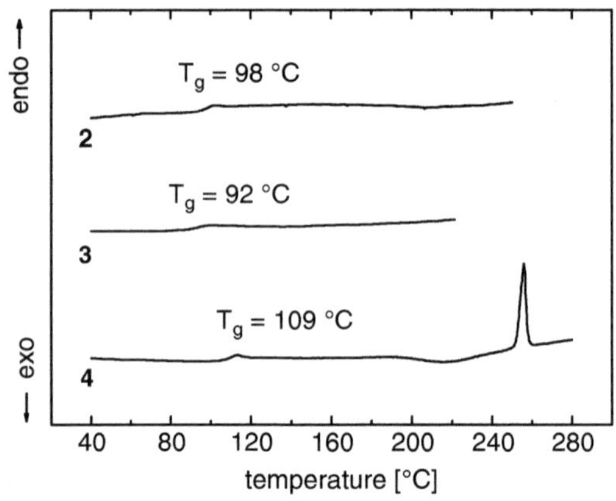

Fig. 11 DSC curves of azobenzene-containing molecular glasses **2**–**4**. Second heating curves at a rate of $10\,K\,min^{-1}$ under N_2

An important aspect of azobenzene-containing molecular glasses is the formation of a stable amorphous phase and of homogeneous, scatter-free thin films with good optically isotropic properties. In the DSC curves in Fig. 11, glass transitions are visible. Compounds **2**–**4** exhibit glass transition temperatures in the range of 92–109°C. Compound **4** recrystallizes before melting, whereas for **2** and **3** no recrystallization could be detected. Compound **3** shows a polymorphic melting behavior which can often be found in molecular glasses. On cooling, no crystallization occurs for all compounds. Therefore, stable amorphous phases can be obtained from all three compounds. The solid-state properties of **2–4** are confirmed by POLMIC investigations and X-ray measurements. From solution, thin amorphous films of these compounds can be spin-coated which remain amorphous if not heated above their glass transition temperature. They showed no crystallization for months when stored at room temperature.

3 Holographic Setup

Azobenzene chromophores are photochromic materials and, therefore, exhibit significantly different absorption spectra for a *trans*-rich state and a *cis*-rich state. Holographic experiments are performed at a wavelength where both species absorb light. In contrast to the writing process, reading of the holographic grating occurs far outside of the absorption region of the chromophores (Fig. 12). In our case the writing laser wavelength is often at 488 nm and the reading laser wavelength at 685 nm.

The holographic setup which was used for the experiments is shown in Fig. 13. Gratings are inscribed with the blue–green (488 nm) or the green line (514 nm) of an Ar^+ ion laser (Coherent Innova 300). A beam splitter divides its output into two coherent beams which are superimposed in the plane of the sample. The beam diameter is about 2 mm. The intensity of each beam is adjusted to $1\,W\,cm^{-2}$. The angle of incidence to the surface normal of the sample is $\pm 14°$, which causes a grating period of $1\,\mu m$. The state of polarization of each writing beam is controlled with a combination of half-wave or quarter-wave plates and Glan–Thomson polarizers. The holograms are read out with an s-polarized red diode laser at 685 nm. In this way the diffraction efficiency can also be monitored in situ during inscription, which is important for determining the rate of hologram formation. The power of the transmitted and the diffracted beam is measured with two photodiodes, the signals of which yield the diffraction efficiency of the grating [72].

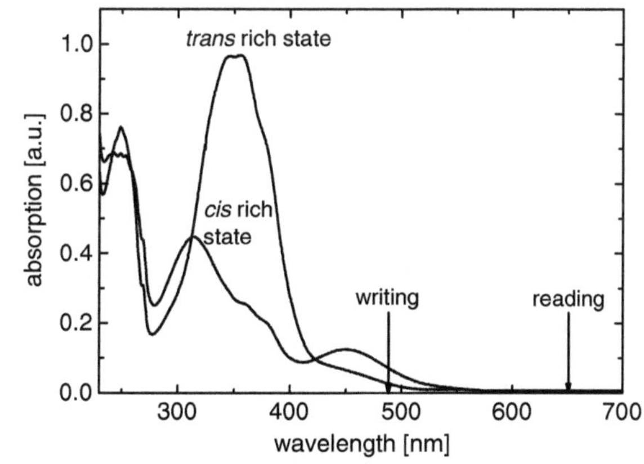

Fig. 12 *cis*-Rich and *trans*-rich absorption spectra of azobenzene derivatives. The writing laser wavelength (488 nm) and the reading laser wavelength (685 nm) are indicated. (Reproduced with permission from [61]. Copyright Wiley-VCH Verlag GmbH & Co. KGaA)

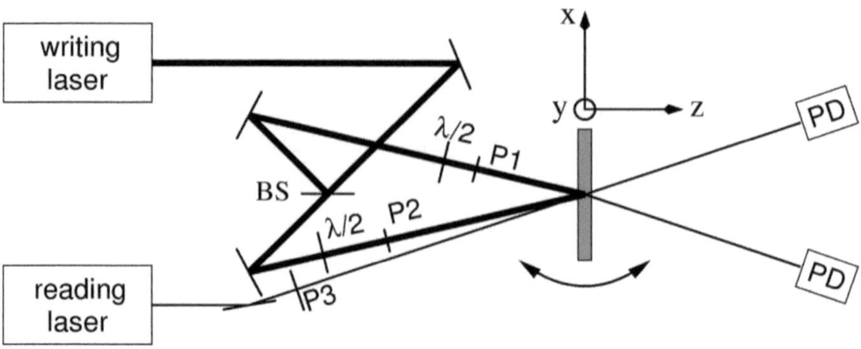

Fig. 13 Schematic representation of the holographic setup. Half-wave ($\lambda/2$) or quarter-wave plates are used to adjust the polarization states of the writing beams. *PD* photodiode; *P*1, *P*2, *P*3 polarizers; *BS* beam splitter. The coordinate system defines the directions which are referred to in the text. (Reprinted with permission from [71]. Copyright 2009 American Chemical Society)

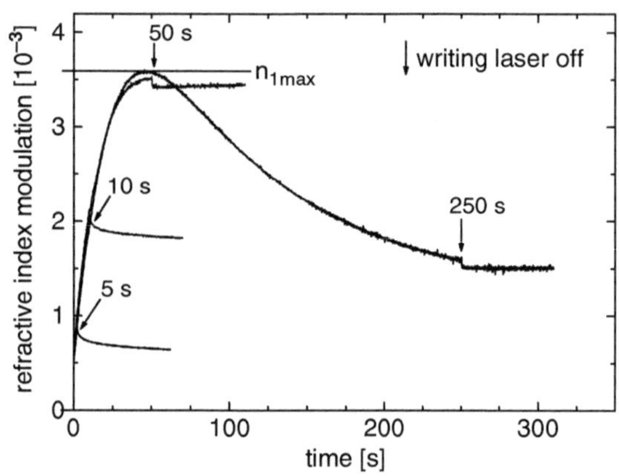

Fig. 14 Typical growth and short-term stability of the refractive index modulation (writing with ss-polarization at different writing times). After 50 s, overexposure leads to a loss of contrast between illuminated and non-illuminated areas

From the measured diffraction efficiency η, it is possible to calculate the refractive index modulation n_1. Assuming the grating vector is oriented perpendicular to the normal of surface of the sample and that the Bragg condition is met, the following formula [73, 74] is valid:

$$\eta = \sin^2\left(\frac{\pi d_0 n_1}{\lambda \cos\theta}\right), \tag{1}$$

where d_0 is the thickness of the sample, λ is the wavelength of the reading beam and θ is its angle of incidence of the readout beam.

A typical course of the refractive index modulation n_1 as a function of time is shown in Fig. 14. The holographic gratings were inscribed in four different experiments for 5, 10, 50, and 250 s. In the beginning, n_1 rises very steeply, until a maximum ($n_{1,\max}$) is reached. Subsequently, here after ~50 s, it decreases again, which is due to a loss of contrast of illuminated and non-illuminated areas. After the writing laser is turned off, a short-term relaxation occurs. The refractive index modulation as a function of time shows a strongly non-linear behavior. In such a case, both the growth during writing and the decay after the laser is turned off can generally be described by stretched-exponential functions [75–78].

For temperature-dependent experiments the sample is placed in a box of anodized aluminum with optical windows. In its walls it contains meandering water pipes connected to an external thermostat (Julabo Ultratemp 2000). Long-term stability experiments up to several years can be performed at room temperature. To this end the samples with inscribed holograms are stored in the dark and their diffraction efficiencies are measured at pre-defined times with a separate diode laser.

The polarization of the laser beams is a highly important parameter, since their combination yields different results during a holographic experiment. The single polarization states are denoted as s (electric field vector perpendicular to the plane of incidence), p (electric field vector parallel to the plane of incidence), 45° (electric field vector at an angle of 45° to the plane of incidence), rcp (right circularly polarized light), and lcp (left circularly polarized light). Their combination yields different polarization configurations, which are summarized in Fig. 15.

Two laser beams with different polarizations can be combined to produce optical gratings. Seven different combinations are mainly used. The configuration sp (s- and p-polarized laser beam brought to interference) generates a pure polarization grating with spatially constant intensity but varying polarization direction. A pure intensity grating is obtained with the combination ss (two s-polarized laser beams). ±45° (+45°- and −45°-polarized laser beam) and "right and left circularly polarized" (rlcp) are mainly polarization gratings with a small contribution of intensity

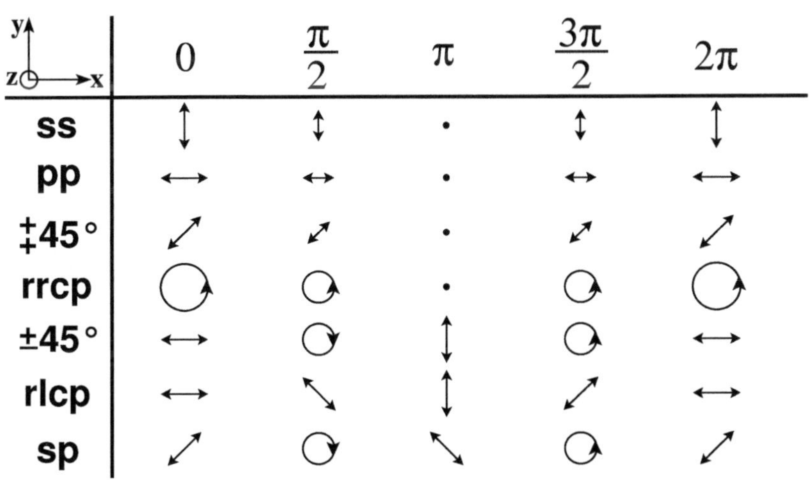

Fig. 15 Electric-field distribution at the sample surface for seven different polarization configurations at five phase differences between the writing beams. (Adapted with permission from [71]. Copyright 2009 American Chemical Society)

variation, whereas pp (two p-polarized laser beams), $++45^\circ$ (two $+45^\circ$-polarized laser beams), and "right and right circularly polarized" (rrcp) are mainly intensity gratings with a slight variation of the polarization direction.

In experiments to produce SRGs, pp, rlcp, and $\pm 45^\circ$ configurations are commonly utilized, whereas the configurations ss and sp are preferred to produce volume gratings. In the case of the volume gratings, polarization gratings as opposed to intensity gratings usually lead to a higher refractive index modulation. Details of the influence of the different polarization configurations on the formation of SRGs are given in Sect. 4.2.3 and [71] and on the formation of volume gratings in Sect. 6.3.1.

As mentioned before, the advantage of holography is that the entire volume of the material can be used for storage and not only the surface. This is realized via multiplexing techniques, e.g., wavelength, shift, phase, or angle multiplexing, and thus it is possible to inscribe several holograms on the same section of the holographic storage material. Angle multiplexing uses the Bragg selectivity of thick holographic grating and, therefore, multiple holograms can be inscribed in the sample at different angles and read out independently.

For writing and reconstruction of complete data pages [79], a separate setup is used which is schematically shown in Fig. 16. Here one of the writing beams (the object beam) is expanded to a diameter of about 5 cm and is sent through a computer-controlled SLM similar to those operating in data projectors. Patterns of dark and bright pixels are generated with a liquid-crystal array. The Fourier transform of the bit pattern (obtained by the focusing achromatic lens behind the SLM) is superimposed with the reference beam on the sample. For reconstruction, the object beam is blocked and the reconstructed light pattern of the diffracted reference beam is imaged onto the sensitive area of a CCD camera (PCO Pixelfly). The arrangement of achromatic lenses between sample and camera warrants that the Fourier transform of the diffracted light pattern can be detected by the camera, which corresponds to the image on the SLM during inscription. Inscribing the Fourier transform in the sample rather than the direct image has the advantage that slight perturbations or

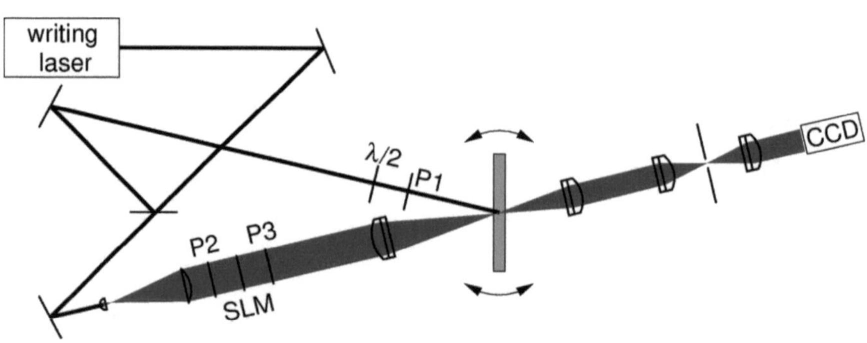

Fig. 16 Setup for simultaneous inscription and reconstruction of entire pages of bit patterns. *SLM* spatial light modulator; *CCD* charge-coupled device camera; *P*1, *P*2, *P*3 polarizers; $\lambda/2$ half-wave plate. (Adapted with permission from [80]. Copyright 2009 SPIE)

imperfections within small area elements of the storage material are less detrimental to the bit error rate [2]. In this setup the reconstruction is performed at the writing wavelength for simplicity, but at a lower intensity.

4 Surface Relief Gratings

Holographic recording materials may develop SRGs in addition to phase gratings in the volume when exposed to interfering laser beams. SRGs obtained from an initially smooth film are associated with macroscopic material transport. In amorphous azobenzene-containing materials, this takes place at temperatures far below the glass transition temperature. This effect was first discovered by Natansohn et al. and Kumar et al. [19, 20] in 1995 and was mainly investigated in polymers carrying azobenzene units as side-groups [81–83].

Understanding the process of SRG formation and finding ways of controlling it is of high technical importance. SRG formation is detrimental to high-density data storage in the volume and must therefore be avoided. On the other hand, there are several applications for which highly uniform SRGs with adjustable spacing and amplitude are of interest. Liquid-crystal anchoring, waveguide couplers, polarization discriminators [84], and antireflective coatings [85, 86] for the visible range are considered as possible applications for SRGs. In addition, SRGs might be used as channel waveguides [87], holographic optical elements (HOE), and in optical security devices [88].

Figure 17 (left) shows a schematic view of a sinusoidal SRG with the grating constant Λ and the modulation height Δd (peak to valley) on a thin film with the initial thickness d_0. The grating period Λ can be adjusted according to the Bragg condition by the angle of incidence of the two laser beams normal to the surface. If films with sufficient thickness are used, the maximum achievable height largely depends on the material. Within this maximum, the height can be easily adjusted by the radiant power. A so-called egg-crate structure (Fig. 17 right) can be obtained if the substrate with a sinusoidal SRG is rotated by 90° and an additional SRG is superimposed. In such a case, the distance from peak to peak corresponds to the

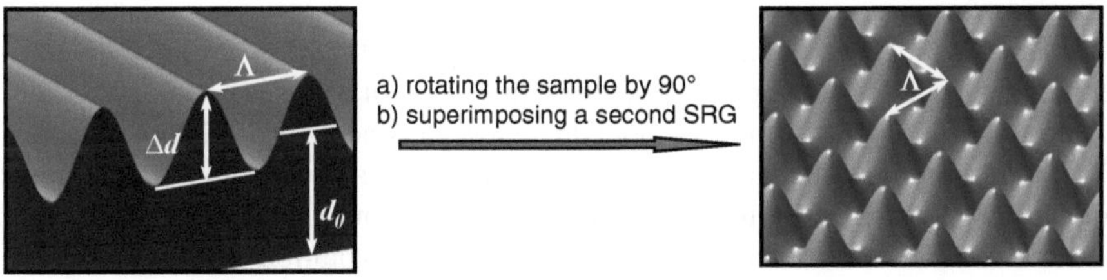

Fig. 17 *Left*: schematic representation of a sinusoidally shaped SRG with the modulation height Δd, initial film thickness d_0, and the grating constant Λ. *Right*: rotating such a sample by 90° and superimposing a second SRG yields an egg-crate structure

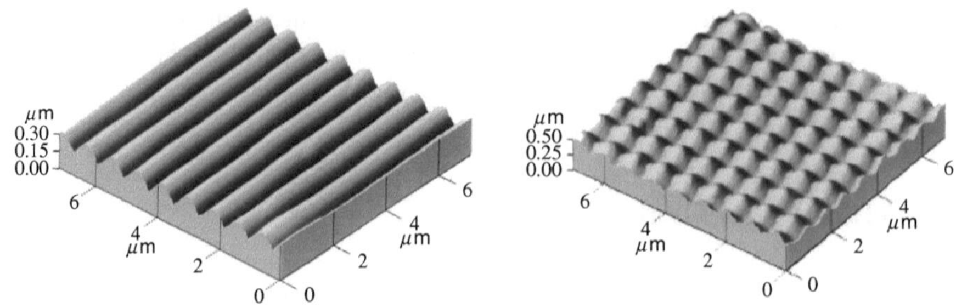

Fig. 18 Inscribed surface structures in an azobenzene-containing polymer; *left*: SRG; *right*: egg-crate structure, which was produced by superimposing two SRGs with the sample being rotated 90° prior to the second inscription process. (Reprinted with permission from [20]. Copyright 1995 American Institute of Physics)

grating constant. In Fig. 18, literature examples of surface structures originated on a thin film of an azobenzene-containing polymer are shown [20]. More complex structures can be found in [84].

It is remarkable that macroscopic material transport takes place more than 100°C below the T_g and that SRGs with grating constants on the order of one micrometer with amplitudes of several hundred nanometers can be formed within a short time. Substantial experimental and theoretical work was devoted to clarifying the mechanism behind SRG formation. A number of theories have been put forth to describe this phenomenon. Hvilsted et al. applied a mean-field model [89] to the formation of SRGs [90]. Rochon and co-workers [91, 92] developed a model in which this effect is ascribed to a pressure gradient between illuminated and non-illuminated areas. The theory of Yager and Barrett [93] explains the SRG build-up by the competition between photoexpansion and photocontraction. Henneberg and co-workers [94] introduced a viscoelastic-flow model. According to the model proposed by Lefin et al. [95], the azobenzene chromophore moves like an inchworm owing to the isomerization between the stretched *trans* and the bent *cis* state, resulting in a translational movement so that the material transport can occur. Although each model describes parts of the observed effects correctly, none of them is able to reproduce all experimental results. Kumar et al. introduced a model based on a time-averaged gradient force. This is the only theory which can account for different polarizations of the laser beams.

Whereas azobenzene-containing polymers in general tend to develop SRGs, we demonstrate in the following that efficient suppression of SRGs is possible, too. This is a basic requirement for holographic data storage, since the SRGs are detrimental to angle multiplexing. In contrast to the suppression, we demonstrate efficient formation of the SRGs by utilizing azobenzene-containing molecular glasses as a novel class of materials. Incipient with the temporal evolution of SRGs, a deeper understanding will be provided by theoretical consideration. This allows one to optimize SRG formation either by the design of the chemical structure of the material or by selecting certain experimental conditions. The formed surface structures are stable enough to be used in a replica molding process and, thus, can be transferred to common polymer surfaces, which opens a variety of further applications.

4.1 Efficient Suppression of SRGs

4.1.1 Azobenzene-Containing Block Copolymers

Efficient suppression of the formation of SRGs must be accomplished if storage in the volume via multiplexing techniques is to be performed. As shown above, it is well known that the illumination of azobenzene-containing materials with holographic gratings causes a fast formation of SRGs. In the following, the SRG formation on solution-processed thin films of different types of azobenzene-functionalized polymers is discussed. We compare the formation of surface gratings between a homopolymer **5**, a statistical copolymer **6**, and a block copolymer **7** functionalized with methoxy azobenzene side-groups (Fig. 19). The syntheses of the polymers were performed as described in Sect. 2.1. A detailed description can be found in [61]. Optical gratings were inscribed with two laser beams in pp-polarization at a wavelength of 488 nm with an intensity of $2\,\mathrm{W\,cm^{-1}}$. After exposure of the surface of the homopolymer **5** with a rather low molecular weight (M_n: $4{,}700\,\mathrm{g\,mol^{-1}}$), a pronounced surface grating with 100 nm modulation height was detected after only 120 s of inscription. The statistical copolymer **6** (M_n: $31{,}000\,\mathrm{g\,mol^{-1}}$; PDI: 1.5, azobenzene content: 29.5 wt%) formed a much shallower SRG of about 5 nm modulation height after an even longer writing time of 250 s. It is still remarkable that, in spite of the relatively low content of photoactive chromophores in the polymer backbone, they are able to move ~70 wt% of the inactive polystyrene material over macroscopic distances. In the case of the film of block copolymer **7** (M_n: $56{,}000\,\mathrm{g\,mol^{-1}}$, PDI: 1.03, 11 wt% azobenzene), absolutely no modification of the surface was detected up to a very long writing time of 3,600 s. This block copolymer is representative of all other block copolymers investigated. In none of them were surface gratings found after a holographic experiment. Macroscopic mass transport is obviously not possible when the azobenzene chromophores are confined in the minority phases and the matrix consists of a polymer block with a glass transition well above room temperature.

4.1.2 Liquid Crystalline Perfluoroalkyl-Azobenzene Polymers

Besides exclusion of the mass transport through confinement in a solid matrix of block copolymers, there are other possible ways to suppress SRG formation.

It is known that a free surface is required for SRG formation. An azobenzene-containing polymer surface covered with a 25 nm-thick non-photoactive layer seems to be sufficient to fully suppress SRG formation [96]. Here, we demonstrate that polymers with azobenzene moieties as side-groups and terminal perfluoroalkyl chains can efficiently suppress SRG formation in a similar manner.

Perfluoroalkyl segments, $-(CF_2)_nF$, are intrinsically liquid-crystalline and can be used as building blocks in self-assembling materials. In combination with other units, such as short alkyl segments, they generate highly organized mesophases [97]. Another important property of the semi-fluorinated moieties is that they can

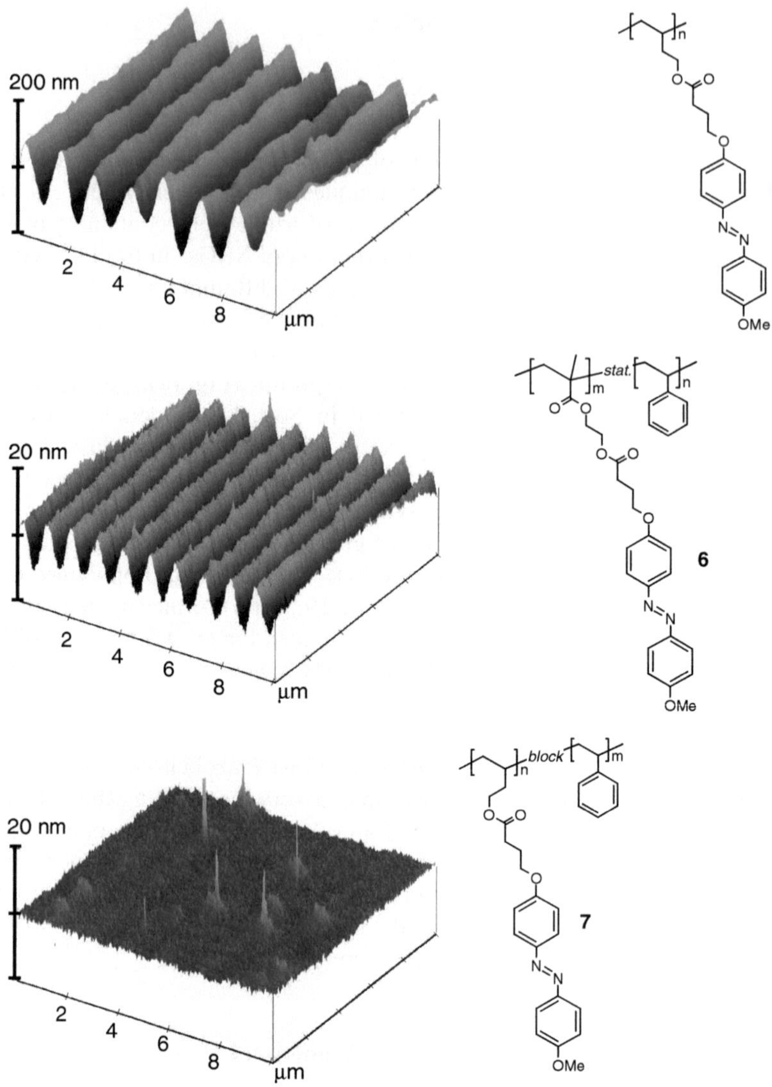

Fig. 19 AFM images of surface relief gratings on thin films of different types of azobenzene-containing polymers. Writing wavelength: 488 nm, intensity: $2\,W\,cm^{-2}$, grating period: $\sim 1\,\mu m$ in all cases. *Top*: homopolymer **5**, writing time: 120 s; *middle*: statistical copolymer **6**, writing time: 250 s; *bottom*: block copolymer **7**, writing time: 3,600 s. Note the different scale bars. (Adapted with permission from [61]. Copyright Wiley-VCH Verlag GmbH & Co. KGaA)

form organized low-energy surfaces [43, 98, 99]. In order to make use of these properties we investigated a series of random copolymers with azobenzene moieties as side-groups and different lengths of the terminal perfluoroalkyl chains **8a–c** (Fig. 20 right). The backbone is based on anionically synthesized polyisoprene with a M_n of $17{,}000\,g\,mol^{-1}$ and a PDI of 1.10. Details of the synthesis of the

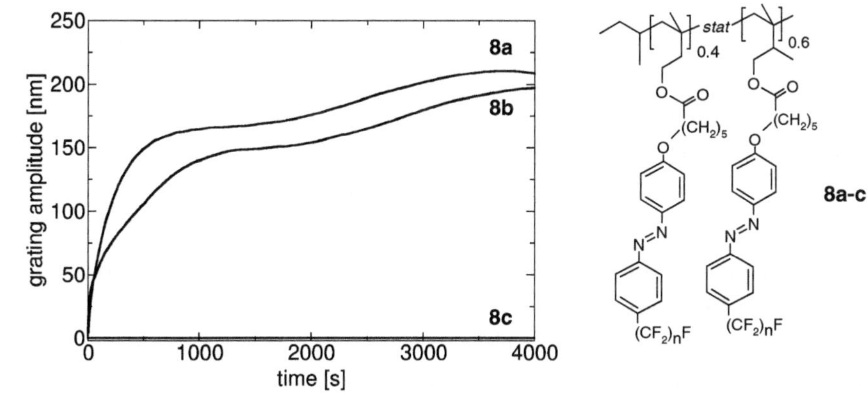

Fig. 20 *Right*: chemical structures of the azobenzene-containing statistical polymers **8a–c** with perfluoroalky segments; the length of the perfluoroalkyl chains is $n = 4$ (**8a**), $n = 6$ (**8b**) and $n = 8$ (**8c**). *Left*: the graph shows the amplitude formation as calculated from the diffraction efficiency according to (2) for **8a**, **8b**, and **8c** vs writing time (writing intensity: $8\,\mathrm{W\,cm^{-2}}$). (Adapted with permission from [100]. Copyright Wiley-VCH Verlag GmbH & Co. KGaA)

perfluoroalkyl polymers are published and can be found in [101]. Liquid-crystalline properties were examined by DSC and wide-angle X-ray scattering (WAXS). For the polymers **8b** ($n = 6$) and **8c** ($n = 8$), a smectic phase was identified with layer spacings of 4.6 and 5.2 nm, respectively. For the polymer **8a** ($n = 4$) no mesophase was observed [100]. From near-edge X-ray absorption fine structure (NEXFAS) analysis and contact-angle measurements it is concluded that in thin films of these polymers the fluoro-groups are on the surface and highly oriented. The order increases with the length of the perfluoroalkyl segments [101].

Holographic experiments were conducted on thin films of the polymers **8a**, **8b**, and **8c** with initially smooth surfaces as previously evidenced by AFM. After exposure, the surfaces were investigated by AFM again. Surprisingly, with a light intensity of $8\,\mathrm{W\,cm^{-2}}$, writing times of several thousand seconds were required to reach the maximum of the diffraction efficiency. Since the diffraction efficiency of SRGs is much larger than the diffraction efficiency arising from the oriented chromophores, the grating amplitude, $\Delta d(t)$, can be directly calculated from the diffraction efficiency according to the equation for thin gratings from Magnusson und Gaylord. It reads

$$\eta = J_1^2\left(\frac{2\pi\Delta d(t)\Delta n}{\lambda\cos\theta}\right) \tag{2}$$

with J_1 being the Bessel function of first order, θ the angle of incidence of the laser beam, λ the vacuum wavelength of the reading laser, and Δn the difference between the refractive indices of the azobenzene-containing material and air. From Fig. 20 it can be seen that the non-liquid crystalline azobenzene polymers **8a** and the smectic polymer **8b** form SRGs under these conditions. In contrast, the polymer **8c**, with the longest chain, did not form SRGs. These results are confirmed by AFM investigations. Nevertheless, the diffraction efficiency for **8c** is not zero and this indicates that a phase grating in the bulk is present instead of an SRG.

If the perfluorinated chain is long enough and smectic phases are formed, thin films of such polymers exhibit a highly rigid and self-assembled liquid crystalline order of the perfluoro-groups on the surface, which prevents the subjacent azobenzene chromophores from forming SRGs. But in-plane orientation in the volume still remains possible. More details can be found in [100].

4.2 Efficient Formation of SRGs with Azobenzene-Containing Molecular Glasses

Besides SRG formation in thin polymer films with azobenzene side-groups, this phenomenon was recently observed in azobenzene-containing molecular glasses [56, 70, 102–104]. In general, molecular glasses are similar to polymers and exhibit a glass transition, which is usually associated with polymers. However, compared to polymers, molecular glasses are well-defined and possess one chemical structure, uniform molecular weight, the absence of structural defects, and the lack of undefined end groups.

The following section is dedicated to the SRG formation in molecular glasses. As mentioned above, numerous models have been proposed to describe the process of SRG formation [90, 92–94, 105, 106], but none of them is able to explain fully all experimental details. In particular, most of the models do not explicitly take the polarization state of the laser beams into account. The only exception is the model by Kumar et al. [107] which relates SRG formation to a gradient force. The following sections are in large parts devoted to experimental and theoretical considerations to validate and enlighten the processes behind SRG formation. The presented investigations were performed on thin films of azobenzene-containing molecular glasses based on a triphenylamine core. The synthesis was performed as described in Sect. 2.2. The chemical structures of the molecular glasses **9a–e** and their glass transition temperatures are shown in Fig. 21. The T_g values vary from 89 to 122°C.

4.2.1 Temporal Evolution of SRG Formation

The time which is necessary to achieve the maximum modulation height is of practical interest. As shown above in (2), for thin gratings the diffraction efficiency can be correlated to the height. Since the reading laser is outside the absorption band, the temporal evolution of the diffraction efficiency and, hence, of the height can be directly monitored without influencing the grating during inscription. Exemplarily, the temporal evolution of the diffraction efficiency on a solution-processed thin film of the triphenylamine-based molecular glass **9b**, which bears a terminal CF_3 substituent, is shown in Fig. 22. Prior to the holographic experiments, the films were annealed at 80°C below the glass transition temperature (T_g of **9b**: 99°C) to remove residual solvent traces. The curve in Fig. 22 reveals that the maximum diffraction

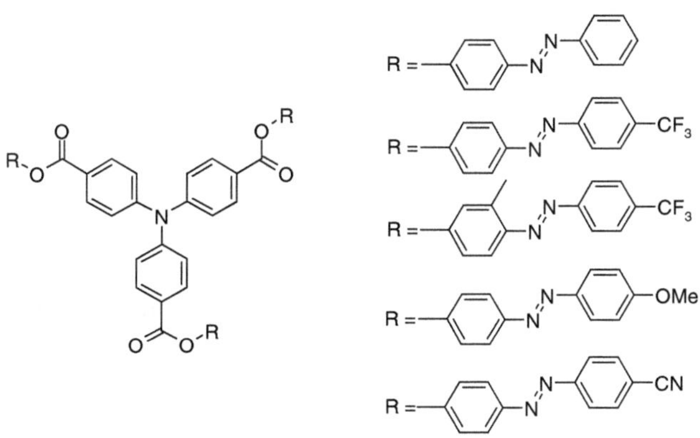

No	T_g
9a	92°C
9b	99°C
9c	93°C
9d	89°C
9e	122°C

Fig. 21 Chemical structures of azobenzene-containing molecular glasses **9a–e** based on triphenylamine and their glass transition temperatures

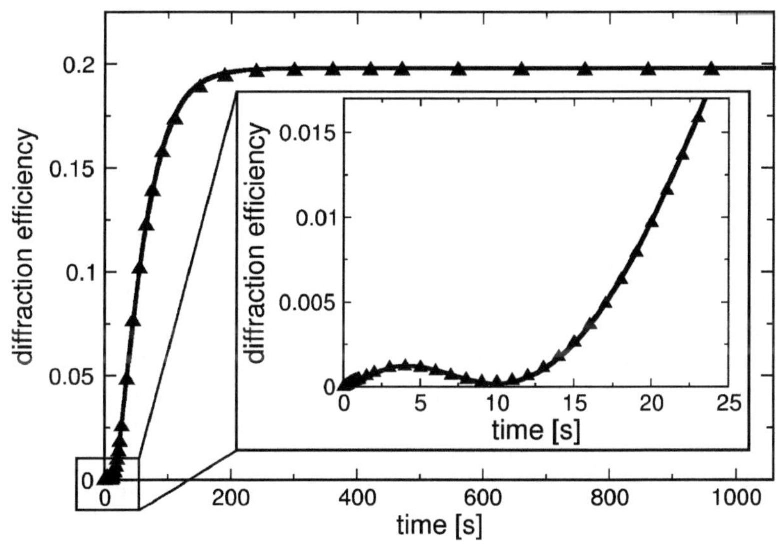

Fig. 22 Temporal evolution of the diffraction efficiency $\eta(t)$ during hologram inscription on a thin film of the azobenzene-containing molecular glass **9b**. Writing was performed with the laser polarization configuration $\pm 45°$. The *black triangles* corresponds to measured data, whereas the *curve* represents the fit of (3). In the first few seconds, the diffraction efficiency due to reorientation of the chromophores in the bulk is visible (see *inset*); later-on the influence of the SRG dominates. (Reprinted with permission from [71]. Copyright 2009 American Chemical Society)

efficiency and, therefore, the height is reached very quickly. The time to reach the maximum in molecular glasses can be in the range of an order of magnitude faster than in polymers. This is representative for all molecular glasses and in agreement with the observations of Kim et al., who attribute the fast and efficient formation to the absence of entanglements in molecular glasses [55, 108].

A closer look at the curve in Fig. 22 (see inset) yields that, within the first few seconds, the diffraction efficiency crosses a small maximum and decays to zero, followed by a steep rise. This indicates that there are two main processes which contribute to the growth of the diffraction efficiency: the volume grating, which is due to chromophore reorientation dominating in the first few seconds, and the SRG, which mostly contributes to the observed diffraction efficiency at later times. Since they are out of phase by 180°, the diffraction efficiency returns to zero after a few seconds. After 60 s of inscription, the diffraction efficiency of the SRG is always much higher than that of the volume grating by at least one order of magnitude. Typical values of the maximum diffraction efficiency of the volume grating in the experiments presented here are around 0.1%, whereas for the SRGs they can reach 30% or more.

In simulations of the temporal evolution of the diffraction efficiency, both of the above processes must be taken into account, as has been described by Reinke et al. [109] and later by Sobolewska et al. [110]. In the model of Sobolewska et al., the time-dependent diffraction efficiency reads

$$\eta(t) = J_1^2 \left\{ \frac{2\pi \left[(n_1(t)d_o)^2 + (\Delta d(t)\Delta n)^2 + 2n_1(t)d_o\Delta d(t)\Delta n \cos(\Delta\phi) \right]^{\frac{1}{2}}}{[\lambda \cos(\theta)]} \right\}, \quad (3)$$

where the first term of the sum describes the increase of the refractive index modulation of the volume grating up to its maximum value $n_{1,\max}$ with an exponential function, the second term describes the growth of the SRG height up to its maximum $\Delta d_{\max}$, and the third term the coupling between the two gratings. τ_1 and τ_2 are the corresponding time constants of the build-up of the volume and SRGs, respectively. Δn is the amplitude of the refractive index change between sample and air, d_0 the thickness of the film, and $\Delta\varphi$ the phase difference between volume and SRG which is assumed to grow to a maximum value $\Delta\varphi_{\max}$ over time. The fit of (3) in Fig. 22 matches the experimental data very well. The maximum phase difference $\Delta\varphi_{\max}$ between the two gratings is obtained as 180°.

With (3) it is possible to calculate the modulation height of an SRG, $\Delta d(t)$, from the measured diffraction efficiency $\eta(t)$ at all times t. Nevertheless, the diffraction efficiency arising from the SRGs is often much higher than that arising from the chromophore reorientation, so that the latter is neglected for high SRGs. By neglecting the second and third terms in the square-root expression of (3) one ends up with the simpler (2). Hence, for high SRGs, both equations yield the same result. The SRG heights calculated with (2) were independently verified by atomic-force microscopy (AFM) measurements. In all cases, the SRG height as measured by AFM and the value calculated from (2) differed by no more than 13%. In Fig. 23 an AFM image of a SRG on a thin film of the methoxy-azobenzene-based molecular glass **9d** is shown. It exhibits a perfectly sinusoidal cross-section with a modulation depth (peak to valley) of 610 nm and a grating constant of 1.0 μm.

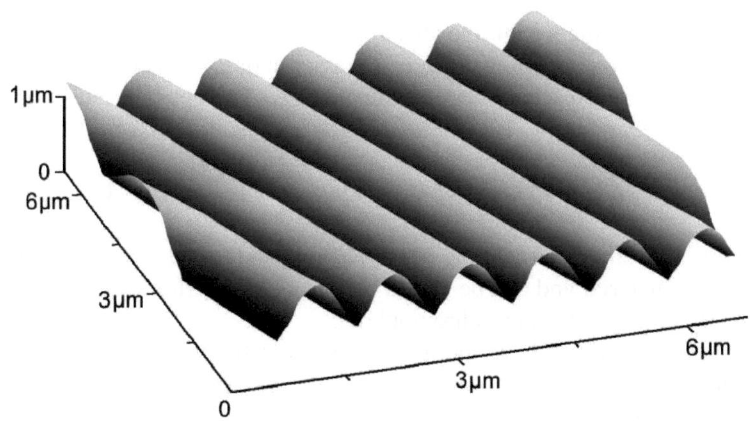

Fig. 23 AFM image of a SRG on a thin film of the azobenzene-containing molecular glass **9d**. The modulation height of the grating is 610 nm with 1.0 µm grating period obtained with a ±45° polarization configuration. (Reprinted with permission from [71]. Copyright 2009 American Chemical Society)

4.2.2 Formation of SRGs: The Gradient Force Model

The reason for this remarkable mass transport during illumination is still not fully understood and several theories have been devised to explain the mechanism behind the formation of SRGs. But as mentioned above, with Kumar's gradient force model it is possible to account for the polarization of the laser beams which obviously plays an important role during the build-up of a SRG. In the framework of this theory [84, 107] the formation of a SRG is ascribed to the presence of a time-averaged gradient force f:

$$\vec{f} = \left\langle \left(\vec{P}\nabla\right)\vec{E} \right\rangle = \left\langle \left(\left(\varepsilon_0 \chi \vec{E}\right)\nabla\right)\vec{E} \right\rangle, \tag{4}$$

where $\vec{P}$ is the polarization induced by the electric light field and χ the electrical susceptibility of the material at the optical frequency of the laser. In the following we refer to the coordinate system as shown in Figs. 13 and 15, where the x axis defines the direction of the grating vector, the z axis is perpendicular to the surface of the sample, and the y axis is perpendicular to the plane of incidence. For reasons of symmetry, macroscopic material transport can only occur parallel to the x axis and, therefore, parallel to the grating vector. In this case, (4) can be simplified to

$$f_x = \varepsilon_0 \left(\chi'_{ix} E_x \frac{\partial}{\partial x} E_x \right), \tag{5}$$

where E_x is the x component of the electric field vector and the subscript i stands for the spatial coordinates x, y, and z. The real part of the optical susceptibility $\chi'_{\mathrm{ix}} = \chi'_{\mathrm{ix,ini}} \pm \Delta\chi'_{\mathrm{ix}}$ is the sum of the initial material susceptibility $\chi'_{\mathrm{ix,ini}}$ prior to inscription and its spatially varying component $\Delta\chi'_{\mathrm{ix}}$ after irradiation. The electric-field vector in the x–y plane is shown in Fig. 15 for different polarizations of the writing beams at several values of the phase difference between them.

Assuming the influence of the electric field to be constant and comparing different azobenzene-containing molecular glasses at one given polarization setting, then only the material susceptibility χ' influences f_x. The optical susceptibility can be written as

$$\chi = (n - ik)^2 - 1 = n^2 - k^2 - 1 - i2nk \quad \rightarrow \quad \chi' = n^2 - k^2 - 1. \qquad (6)$$

The absorption coefficient k and the refractive index n (and their spectral variation) depend on the material and can be varied by introducing different substituents into the periphery of the azobenzene chromophore.

Volume phase gratings have the minima of the refractive index modulation in the illuminated regions and the maxima in the dark regions of the material, because the long axis of the shape-anisotropic azobenzene moieties (with the highest polarizability) is turned away from the electric-field vector. For a phase shift of 180° between SRG and volume grating, the maxima of the SRGs must then be in the illuminated areas. Hence, the direction of material transport of the SRGs must be from dark to light regions, which is in accordance with (5) for positive χ'_{ix}, as is the case for the investigated materials.

4.2.3 Influence of the Polarization of the Writing Beams

On the basis of the above-mentioned theoretical considerations, holographic experiments with different polarization settings of the writing beams were performed. These led to different SRG heights, which can be explained by the different strengths of the gradient force f_x. The term $E_x(\partial/\partial x\, E_x)$ can be written as the product of a periodic function of x, which is the same in all seven settings, and an amplitude which depends on the polarization. The normalized values of f_x are summarized in Table 3.

In ss and sp configuration, either the x component of the electric field vector E_x or its derivative is zero, so f_x is zero and no significant SRG is observed.

Table 3 Comparison of selected characteristics of the SRGs for different polarization states of the writing beams

Polarization configuration	ss	pp[a]	++45°	rrcp	±45°	rlcp	sp
Normalized f_x	0	1	0.5	0.5	0.5	0.5	0
Normalized average maximum height of SRG[b]	0.05	1	0.83	0.92	0.77	0.77	0.12
Normalized average height of SRG after 60 s[b]	0.09	1	0.38	0.41	0.96	0.93	0.07
Maximum growth rate of **9d** ($\mathrm{nm\,s^{-1}}$)	0.9	17.7	10.6	9.5	21.2	20.7	1.3
Time to reach maximum growth rate of **9d** (s)	67	29	49	49	16	16	70

[a] For all normalized quantities, the value of pp has been arbitrarily set to 1

[b] Heights are given as an average of all materials with respect to the corresponding polarization configuration

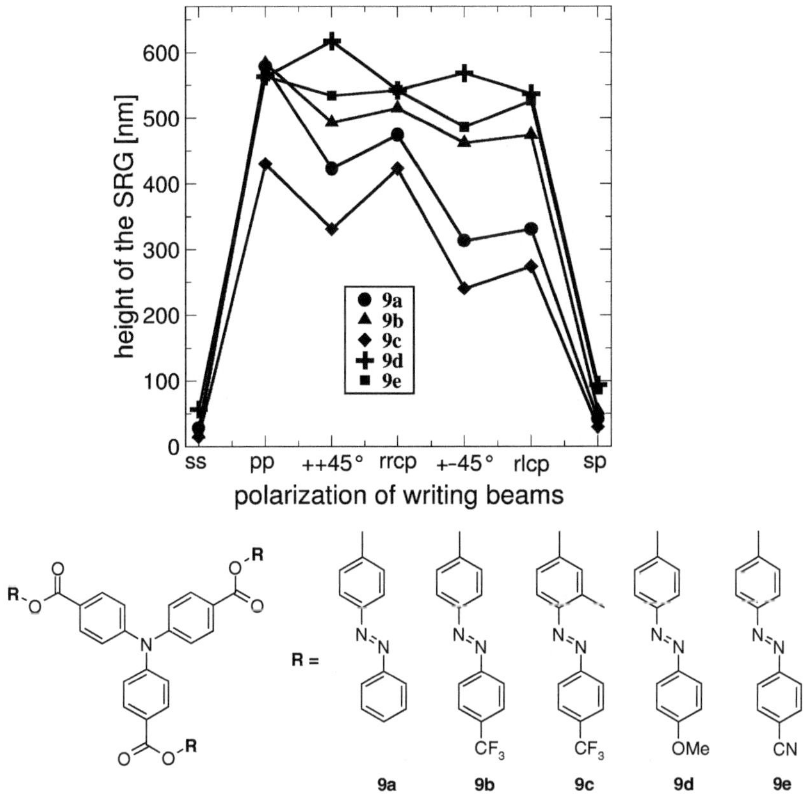

Fig. 24 Maximum height of the SRGs of azobenzene-containing molecular glass series **9a–e** as calculated from the measured diffraction efficiency (according to (2)) for the different polarization settings. The *lines* are a guide to the eye. (Adapted with permission from [71]. Copyright 2009 American Chemical Society)

The polarization settings $\pm45°, ++45°$, rlcp, and rrcp have the same normalized $f_x = 0.5$; hence, they generate comparable SRG heights as can be seen in Table 3 and in Fig. 24. The highest value of f_x is obtained with two p-polarized writing beams. Consequently, this configuration yields the highest SRGs of all polarizations. It is reasonable to assume that the maximum SRG height is reached when f_x is balanced by the surface energy. The fact that the height obtained with pp is not twice that of, for example, rlcp indicates that the surface energy varies with the height in a nonlinear fashion.

The SRG heights after 60 s of illumination are given in Table 3. They behave differently compared to the maximum heights shown in Fig. 24. After 60 s, the settings $\pm45°$ and rlcp generate much higher SRGs than $++45°$ and rrcp, although they all have the same value of f_x. According to Kumar et al., an important factor for the build-up of SRGs is the plastification or softening of the material caused by the *trans–cis–trans* isomerization cycles. In the case of the polarization gratings generated by $\pm45°$ and rlcp, such a plastification takes place throughout the material, in contrast to the intensity gratings (++45° and rrcp), which feature no plastification in the dark regions. As a consequence, the SRGs increase more rapidly for the polarization gratings. The maximum height, on the other hand, is slightly higher for the intensity gratings. This may be due to a nonlinear relationship between light intensity and the degree of plastification. Further details can be found in [71].

4.2.4 Influence of the Substituents at the Azobenzene Unit

For a comparison of azobenzene-containing molecular glasses three molecular parameters are important: absorption coefficient, susceptibility, and volume of the chromophore. The glass transition temperature T_g of the investigated materials has no measurable influence on the formation of SRGs. The T_g values of all materials differ by only 30°C and are well above room temperature, e.g., the lowest T_g was determinded to 89°C for compound **9d**.

The maximum SRG heights are shown in Fig. 24. The highest SRGs are formed in **9d**. In **9c**, which shows the lowest values, the formation of SRGs is hindered by the methyl group. The optical densities per micrometer at the writing wavelength of 488 nm, which are proportional to the absorption coefficients, are listed in Table 4.

At a writing wavelength of 488 nm one is still exciting the nπ* transition (see Fig. 12). Therefore, the absorption coefficient is not only influenced by the oscillator strength of the chromophores, but also by slight spectral shifts of the transition due to the different absorption of the *trans* and *cis* states. Stronger absorption leads to the enhanced *trans–cis–trans* isomerization cycles and, therefore, to a stronger plastification of the material at a local scale, yielding higher SRGs.

The absorption constant k and the refractive index n are related via the Kramers–Kronig relations. Hence, an increase of the optical density, either due to an increase of the oscillator strength or a spectral red shift of the transition, leads to an increase of the refractive index. Since n is larger than k by two orders of magnitude and usually decreases more slowly with increasing distance from the absorption maximum, it yields the main contribution to the optical susceptibility. Therefore, the effect of

Table 4 Parameters of SRG formation of the five molecular glass formers

Molecular glass	**9a**	**9b**	**9c**	**9d**[a]	**9e**
Normalized OD/μm at 488 nm	0.71	0.82	0.85	1	1.04
Normalized χ' at 488 nm	0.69	0.79	0.76	1	1.03
Normalized average maximum height of SRG[b]	0.73	0.84	0.58	1	0.89
Normalized average height of SRG after 60 s[b]	0.48	0.48	0.21	1	0.66
Average writing time (s)[b]	201	272	356	87	121
(Normalized average writing time)$^{-1}$ (s^{-1})[b]	0.43	0.32	0.18	1	0.72
Maximum growth rate with pp (nm s^{-1})	7.9	7.2	4.2	17	11
Time to reach maximum growth rate with pp (s)	40	47	64	27	40

[a] For all normalized quantities, the value of **9d** has been arbitrarily set to 1
[b] Heights and writing times for each material are an average of the polarization configurations

plastification due to the higher absorption coefficient is enhanced by an increase of the susceptibility, resulting in higher SRGs. The relative susceptibilities of the five materials as obtained from ellipsometry, which determines the driving force f_x for material transport, are listed in Table 4. They show clear correlations with the maximum SRG height.

4.2.5 Replica Molding of SRGs and Imprinting on Polymer Surfaces

Important technical applications of SRGs are expected in the field of HOEs, security devices, or for adjusting the haptic properties of surfaces. Advantageously, surface structuring by this method is an optical, non-contact approach where no mechanical or chemical processes are required. Superimposition of SRGs allows one to create a variety of complex patterns [84]. As shown earlier, azobenzene-containing molecular glasses can develop SRGs several hundred nanometers in size within minutes. For applications, however, it is convenient for the created surface patterns to be transferred to more stable and durable polymers. But such structures cannot be directly imprinted onto a polymer surface since the polymer must be heated to temperatures close to, or above, the T_g of the molecular glass.

Replica molding can be used to circumvent this problem. For this, a liquid pre-polymer is cast over the master (the molecular glass) which has the holographic relief structure on its surface. As pre-polymer, the commercially available PDMS silicone elastomer (Sylgard 184, DOW Chemical) was used, which is characterized by good thermal stability, optical transparency, and low interfacial free energy. The SRGs on the glass film are stable enough for replicas to be cast, which are then, after thermal curing, used as dies for imprinting onto polymer surfaces. Such a process sequence is shown in Fig. 25. The SRG master was inscribed on a film of azobenzene-containing molecular glass **9e**. With a glass transition temperature of 122°C, the SRGs resist in the curing temperature of PDMS of 60°C (step I). After separating it from the master, its surface structure was imprinted onto a polycarbonate sheet in a heated laboratory press (heated to 180°C and pressed with a force of 5 kN for 2 min; step II). All steps were characterized by AFM. The results of the AFM measurements are presented in Fig. 25 for a typical example. All three gratings (master, replica, and imprint) show only slight distortions.

5 Holographic Volume Gratings

Holographic volume gratings in homopolymers and statistical copolymers with azobenzene side-groups have been investigated by several authors [10–14, 16, 112]. In general, these materials develop not only phase gratings but also SRGs when illuminated with polarized light. For data storage applications, where multiple holograms are to be inscribed at the same position by multiplexing techniques, the latter must be avoided. It is indeed possible to inscribe simple volume gratings

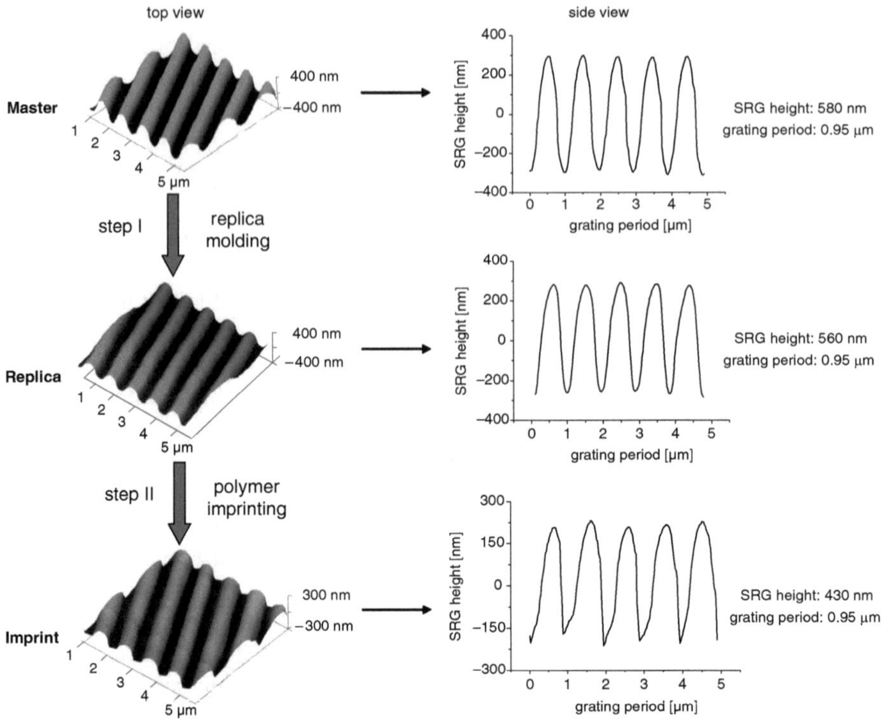

Fig. 25 Example of replica molding and imprinting of a surface pattern. *Left*: 3D surface plots of AFM measurements of the original SRG structure generated holographically on compound **9e** *(top)*, the replica on the silicone elastomer *(middle)*, and a polycarbonate imprint *(bottom)*. *Right*: corresponding cross-sections. (Adapted with permission from [111]. Copyright Wiley-VCH Verlag GmbH & Co. KGaA)

in homopolymers and copolymers. Our approach is to use diblock copolymers as storage materials, since they exhibit marked benefits as compared to homopolymers and copolymers. A comparison of azobenzene-containing homopolymers, statistical copolymers, and block copolymers with respect to the photophysical properties of their volume gratings, such as achievable refractive index modulation as well as short-term and long-term stability, will be presented in the following to reveal the advantages of the block copolymer approach. In particular, structure-property relations will be given and a materials design will be discussed.

5.1 Inscription and Evaluation of Volume Phase Gratings in Thin Films

The inscription of a single holographic volume grating was performed on homogeneous transparent thin films. Particularly in homopolymers and copolymers with a high azobenzene chromophore content, the film thicknesses are limited to about 100 μm due to their high optical densities, which should not exceed ~0.7 at the writing wavelength in order for the light to penetrate the entire sample.

Polymeric thin films in this thickness range can easily be prepared from solution by spin coating or doctor blading techniques. Drying the films at elevated temperatures under high vacuum or under inert atmosphere is necessary to remove residual solvent traces, which cause softening of the material and contribute to fast relaxation of the oriented chromophores to an isotropic state.

To compare different materials and preparation conditions, holographic gratings were always inscribed up to the first maximum of their diffraction efficiency η which – as was shown in Sect. 3 – is connected with the refractive index modulation n_1. The refractive index modulation was calculated from the diffraction efficiency according to (1). Besides the value of the refractive index modulation, the stability of the grating is another important parameter which can be observed by monitoring the diffraction efficiency for a given time. A further parameter which is related to the refractive index modulation, and therefore to the diffraction efficiency, is the sensitivity. The sensitivity can be considered to be the inscribed refractive index modulation per time and is defined as

$$S'(t) = \frac{\sqrt{\eta}}{Itd} \tag{7}$$

with I denoting the intensity of the laser beams, t the writing time, and d the film thickness. For a holographic data storage material a high sensitivity is desirable. One has to note that there is more than one possible expression for the calculation of the sensitivity of a holographic storage medium. However, in each equation the light intensity and the writing time are inversely proportional to the sensitivity. Hence, a highly sensitive material is obtained if less radiant energy is needed for a specific refractive index modulation. But in general a fair comparison between material classes can hardly be made. In an irreversible photopolymer system, the light energy is only necessary for the initiation of the polymerization and not during the whole grating formation process, which happens due to the diffusion of the monomers after irradiation. In contrast, azobenzene-containing materials are irradiated throughout the whole grating formation process and, therefore, yield a low sensitivity in comparison to photopolymer systems.

All calculations from the measured diffraction efficiency to other numbers characterizing the material require the knowledge of either the sample thickness or the angular dependence of the diffraction efficiency. We chose the latter method since angle-dependent measurements can be performed easily in our set-up. The thickness values calculated from the holographic data were compared with independent measurements performed with a step profiler. The results always agreed within a few percent. To calculate n_1 we inserted the optically determined thickness, since it corresponds exactly to the location of the hologram.

5.2 *Improvement of Refractive Index Modulation and Stability*

5.2.1 Comparison of Homopolymers, Statsitical Copolymers, and Block Copolymers

The refractive index modulation of a holographic grating determines the fraction of the intensity of the laser beam which is diffracted. If hundreds of holographic gratings are to be superimposed at the same spot, all gratings have to share this value. It is obvious that the achievable refractive index modulation should be as high as possible in order to inscribe as many holograms as possible in a defined volume. In azobenzene-containing materials the refractive index modulation often reaches values in the range of 10^{-4} to 10^{-2}, depending on the azobenzene content in the polymer film and on the chemical structure of the chromophore. In particular, azobenzene side-groups forming liquid crystalline mesophases generally exhibit far better refractive index modulations. As described in Sect. 2.1, nematic mesophases are formed in cyano-azobenzene-containing side-group polymers. As we show in the following, the achievable refractive index modulation also depends on the polymer type. This is illustrated in Fig. 26 where the values of a homopolymer, a statistical copolymer, and block copolymers **1a–e** (see Table 1) based on cyano-azobenzene side-groups are compared. The homopolymer, which has the highest concentration of azobenzene moieties, shows the highest refractive index modulation $n_{1,\max}$ with a value of 1.6×10^{-2}. For the block copolymers **1a–e** the n_1 values increase with increasing azobenzene block length, i.e., with higher azobenzene concentration. The statistical copolymer exhibits a refractive index modulation of only 1.2×10^{-3}, which is more than an order of magnitude lower than for the

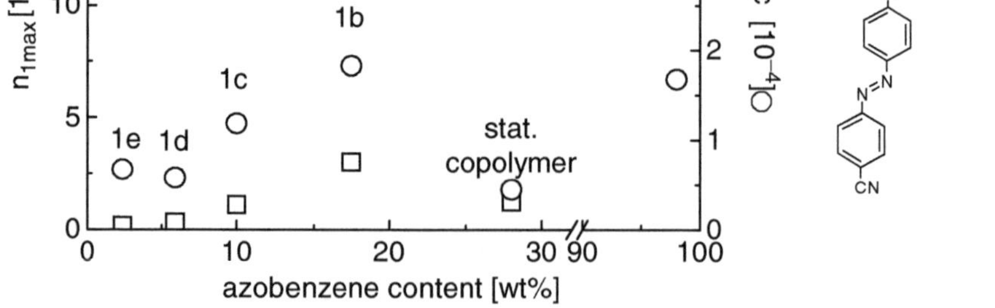

Fig. 26 Maximum refractive index modulation ($n_{1\,\max}$) and azobenzene-content normalized maximum refractive index modulation ($n_{1\,\max}/c$) of different types of cyano-azobenzene-containing polymers

homopolymer. It should be pointed out that at a similar azobenzene concentration, block copolymer **1a** has a much higher maximum refractive index modulation than a statistical copolymer. This strongly supports the concept of block copolymers as ideal materials for further development of holographic storage media. The main difference and advantage with respect to the statistical copolymers utilized here is the confinement of the azobenzene side-groups in the nanostructured morphology. The improvement is even more pronounced when the refractive index modulation is normalized with respect to the chromophore concentration. All the block copolymers have higher values of $n_{1,max}/c$ than the statistical copolymer and the block copolymers **1a** and **1b** have even higher values than the homopolymer.

As mentioned, within the block copolymer series **1a–e** the refractive index modulation $n_{1,\max}$ decreases with decreasing chromophore content. Interestingly, the same trend can be observed in the normalized values. One might assume, therefore, that the refractive index modulation is correlated with the block length of the shorter blocks, since these steadily decrease from **1a** to **1e** (Table 1).

Naturally, block copolymers exhibit different morphological phases if the volume ratio of the two blocks is changed. The block copolymers **1a** and **1b** form cylindrical morphologies, while for **1c** and **1d** the azobenzene side-groups are confined in spheres, but exhibit different azobenzene contents and, therefore, different domain sizes. From the figure one cannot clearly conclude that the domain size plays a role for the maximum achievable refractive index modulation. Thus – as we will demonstrate in the following section – one has to investigate the influence of the morphology.

Besides the absolute value of the refractive index modulation, the stability is an important parameter, particularly for the development of a storage material. In Fig. 27 the temporal variation of the diffraction efficiency of volume gratings is plotted for films of a statistical copolymer **6** (azobenzene content: 29.5 wt%) and

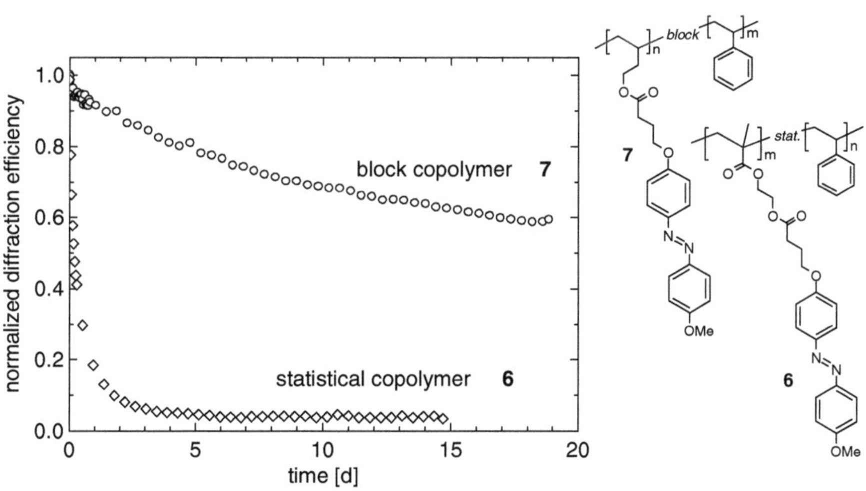

Fig. 27 Stability of volume gratings in thin films of the statistical copolymer **6** and block copolymer **7** with methoxy-azobenzene side-groups

a block copolymer **7** (11 wt% azobenzene), both containing methoxy-azobenzene side-groups and a comparable azobenzene content. The stability is quite different. Whereas more than 60% of the initial diffraction efficiency is still present in the block copolymer after 2 weeks, it decays nearly completely within a few days in the statistical copolymer. Here again the confinement and the cooperative effect of the azobenzene side-groups, which are present in the block copolymer, are responsible for this improvement.

5.2.2 Influence of the Block Copolymer Morphology

It was demonstrated above that both the value and the stability of the refractive index modulation are far better for block copolymers than for statistical copolymers with comparable azobenzene content. But the influence of the morphology and the domain size on the refractive index modulation remains unclear. To investigate this we chose a block copolymer series with three different morphologies: lamellar, cylindrical, and spherical. In Fig. 28 the growth and decay for different morphologies of the block copolymers **10** (M_n: 45,900 g mol^{-1}; azobenzene content: 31 wt%) and **11** (M_n: 42,400 g mol^{-1}; azobenzene content: 25 wt%), which are based on a PMMA–PHEMA backbone, are shown [66, 113]. Both block copolymers exhibit a similar

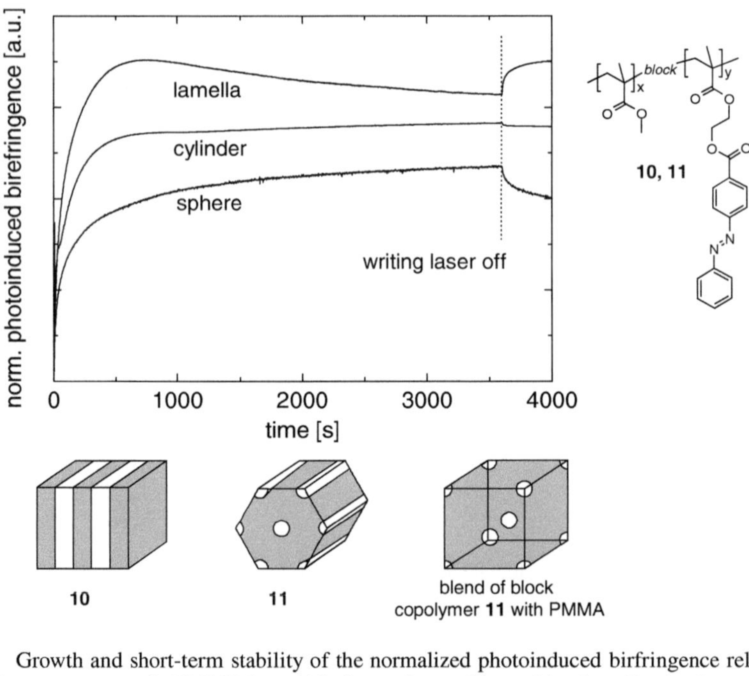

Fig. 28 Growth and short-term stability of the normalized photoinduced birfringence relating to the azobenzene content in PMMA-based block copolymer films with a lamellar (azobenzene content: 31.6 wt%), cylindrical (23 wt%), and spherical (7.9 wt%) morphology. Writing time: 3,600 s, writing wavelength: 532 nm, reading wavelength: 630 nm

molecular weight, but **10** forms a lamellar morphology, whereas **11** forms a cylindrical morphology. In order to obtain a spherical morphology it is possible to blend block copolymer **11** with a PMMA matrix homopolymer, until a transition from the cylindrical morphology to a spherical one occurs. In all three morphologies, the azobenzene chromophores are reoriented in their constrained environment upon irradiation with polarized light, but great differences in behavior during the writing process as well as during the decay of the gratings are observed. The maximum induced birefringence normalized to the azobenzene concentration was found to be different and depends on the morphology. The highest photoinduced birefringence was measured for the lamellar morphology and the lowest for the spherical morphology. In the lamellar structure, the maximum birefringence was reached after less than 800 s and then decreased, whereas in the morphologies with cylinders and spheres the birefringence continued to increase slightly. After the laser is turned off, a significantly different short-term relaxation behavior is observed. The spheres show only a slight decay, the cylindrical morphology has nearly a constant value, and the lamellar phase even shows an increase in the refractive index modulation. One has to note that the diameters of the cylindrical and spherical morphologies are the same (12 nm as determined by TEM), since they are based on the same block copolymer and, therefore, have the same length of the shorter block. However, they behave differently in the holographic experiments, and because of this we have to conclude that the performance of thin films of azobenzene-containing block copolymers during a holographic experiment depends on the block length, the morphology, and the domain size.

5.2.3 Influence of Azobenzene Side-Groups in Block Copolymers

It is known from the literature that the chemical structure of the azobenzene side-groups influences the photophysical properties. In particular, liquid crystalline moieties improve the refractive index modulation. In block copolymers, however, the influence of the chromophores is difficult to assess due to the additional dependence of the refractive index modulation on morphology and domain size. Therefore, a series of block copolymers with different azobenzene chromophores but similar backbones was investigated. The data are normalized to the azobenzene content for better comparison. The investigated block copolymers series is based on a polystyrene–polybutadiene backbone. The chemical structures are shown in Fig. 29 and the characteristic data on the molecular weight, PDI, and azobenzene content are summarized in Table 5. Volume gratings were inscribed in thin solution-processed films with ss polarization of the laser beams (writing wavelength: 514 nm; fluence: 500 mW cm^{-2}). Three main parameters were investigated: the refractive index modulation, the stability of the volume gratings, and the sensitivity S'. In azobenzene-containing materials, the refractive index modulation and, therefore, the diffraction efficiency and the sensitivity exhibit a non-linear evolution with time. To illustrate this, the sensitivity values of each grating were calculated at the time t for which the refractive index modulation had reached half its maximum value.

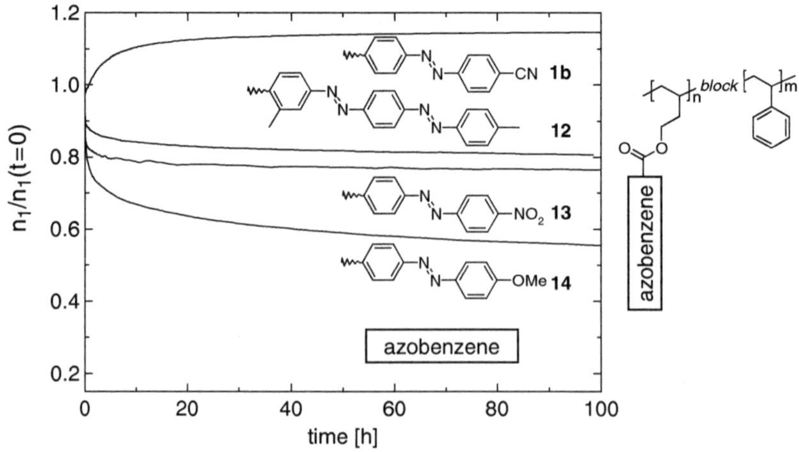

Fig. 29 Long-term stability of holographic gratings in thin films of block copolymers with different azobenzene side-groups

Table 5 Characteristic properties of block copolymers with different azobenzene side-groups

Block copolymer	**1b**	**12**	**13**	**14**
$M_n (g\,mol^{-1})^a$	68,000	71,000	76,000	56,000
PDI	1.15	1.14	1.07	1.03
Azobenzene content (wt%)	17.5	18.9[b]	26.9	10.6
Normalized refractive index modulation $(10^{-2})^c$	2.52	2.46	1.80	1.39
Norm. refractive index modulation after 90 h	1.15	0.81	0.77	0.56
Normalized sensitivity $(10^{-2}\,cm^2\,J^{-1})^d$	4.36	2.01	2.84	1.62

[a] Obtained from SEC with respect to polystyrene standards
[b] Polymeranalogous reaction not quantitatively converted
[c] Normalized to the azobenzene content $(n_{1,max}/c)$
[d] Calculated according to (7); normalized to a thickness with an OD of 0.5 at the writing wavelength

Equation (7) indicates that the sensitivity depends on the sample thickness. Thus, the values of the sensitivity were additionally normalized to a film thickness for which the optical density at the writing wavelength would be 0.5. The results are summarized in Table 5. The normalized long-term stabilities of the gratings are shown in Fig. 29.

The refractive index modulation of the block copolymer **1b** carrying cyano-azobenzene side-groups is more than twice that of the methoxy-substituted block copolymer **14**. In addition, **1b** exhibits a much higher sensitivity than **14**. This is just a consequence of (7), in which the sensitivity increases with the refractive index modulation. Furthermore, **14** exhibits a continuous decay after the writing laser is turned off, similar to that shown in Fig. 27. Block copolymer **1b**, in contrast, even shows an increase by 15% within a few hours, and the refractive index modulation is stable for years. In this case, the so-called post-development is not due to a different morphology as discussed above. Rather, the liquid crystalline character of the chromophores improves the performance by stabilizing the order parameter of the oriented domains after the laser is turned off. Block copolymers **12** and **13** with the bisazo- and nitro-substituted azobenzene, respectively, exhibit a behavior of refractive index modulation and stability which is in between that of the materials with the cyano- and methoxy-substituted azobenzene side-groups.

5.2.4 Block Copolymers with Photoactive Chromophores and Non-photoactive Mesogenic Side-Groups

In Sect. 5.2.3 it was demonstrated that block copolymers with liquid crystalline azobenzene side-groups form stable volume gratings, whereas non-liquid crystalline chromophores do not. This observation was extended to inscribe stable volume gratings with non-liquid crystalline azobenzene chromophores in combination with non-photoactive anisotropic mesogens in the shorter block. This is advantageous for two reasons. First, the cyano-azobenzene chromophore shows a slightly decreasing diffraction efficiency after several inscription processes, whereas the methoxy-substituted chromophore does not. The second reason is that, in this way, the chromophore content can be reduced even further without a loss of refractive index modulation, which is desirable to prepare thick samples for angle multiplexing.

For a detailed investigation, a series of different block copolymers was synthesized which differed with respect to the ratio of azobenzene vs non-photoactive mesogenic side-groups. The introduction of both side-groups is possible by a polymeranalogous reaction and yields a random distribution of azobenzene chromophores and mesogenic side-groups along the backbone of the shorter block. The characteristic data are given in Table 6.

The mesogen content in the photo-addressable minority block varied from 0 (block copolymer **14**) to 76 mol% (**14d**). Thin films of each material were prepared by spin-coating and dried under high vacuum at room temperature or annealed at 50, 80, and 120°C for 20 h. Figure 30 (top) shows the maximum achievable refractive index modulation $n_{1,\mathrm{max}}$ for these block copolymers as a function of mesogen content and annealing temperature. The corresponding writing times of the refractive index modulation are shown in Fig. 30 (bottom).

The achievable refractive index modulation increases with mesogen concentration. This clearly demonstrates that the non-absorbing mesogen side-groups are reoriented jointly with the azobenzene side-groups, i.e., a cooperative effect between the different aromatic side-groups in the minority phases exists. The increase of the $n_{1,\mathrm{max}}$ values with annealing temperature to a maximum can be interpreted in terms of an induced liquid crystalline ordering of the side-groups. With increasing annealing temperature, azobenzene and mesogenic side-groups are expected preferentially to attain parallel orientation, which favors the cooperative reorientation during hologram inscription. If the annealing temperature is too high and approaches the T_g of the matrix, the light-induced orientation of the azobenzene units is no longer stable and, therefore, the maximum refractive index modulation decreases. This can be seen for the samples annealed at 120°C.

Table 6 Characteristic properties of the block copolymer series based on a polystyrene block and a 1,2-polybutadiene block containing methoxy-azobenzene and mesogenic side-groups

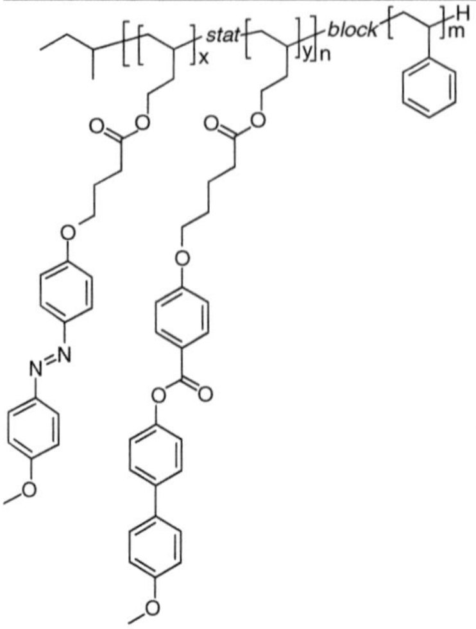

	M_n ($g\,mol^{-1}$)	PDI	w_{PS} (wt%)	w_{PB} (wt%)	Azo content (mol%)	Mesogen content (mol%)
14	56,000	1.03	89.4	10.6	100	0
14a	53,700	1.02	89.2	10.8	86	14
14b	54,100	1.02	89.9	10.1	65	35
14c	54,600	1.02	87.9	12.1	48	52
14d	55,300	1.03	87.4	12.6	24	72
15	65,500	1.03	70.8	29.2	50	50

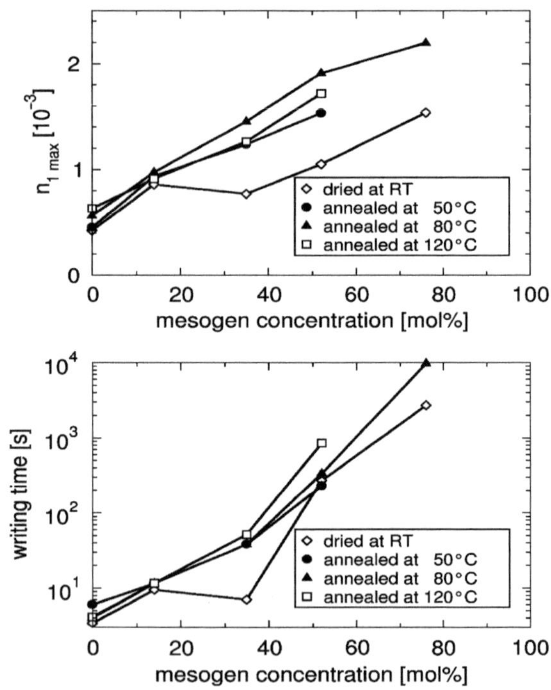

Fig. 30 *Top*: maximum refractive index modulation for the block copolymers **14–14d** after different annealing conditions. In all cases the annealing time was 20 h. Writing wavelength: 488 nm; writing intensity: $2\,W\,cm^{-2}$. *Bottom*: required writing times to the maximum of the refractive index modulation. Note the logarithmic scale in the lower part. (Adapted with permission from [72]. Copyright Wiley-VCH Verlag GmbH & Co. KGaA)

A disadvantage of introducing mesogenic side-groups is that they require much longer writing times, which is unfavorable for data storage applications. Apparently, the long mesogenic groups do not have sufficient free volume in their environment, so their reorientation takes a longer time.

The stability of inscribed holographic gratings was investigated on the four block copolymers **14–14c** with different contents of mesogenic side-groups and the same block length, and on block copolymer **15** with a longer photoaddressable block. The gratings were written up to the maximum of the growth curve. The samples were placed in separate boxes in the dark at a constant temperature of 22°C. Over a period of nearly 2 years, the diffraction efficiencies were measured with a diode laser at pre-defined times. The results are displayed in Fig. 31.

Block copolymer **14** without mesogenic side-groups showed a fast decay of the refractive index modulation already in the first few minutes and had reached half of its original n_1 value after roughly one week (Fig. 31). As little as 14 mol% mesogenic side-groups (**14a**) led to much improved stability; the hologram decayed by only 18% during a period of 270 days. In the case of block copolymers **14b**,

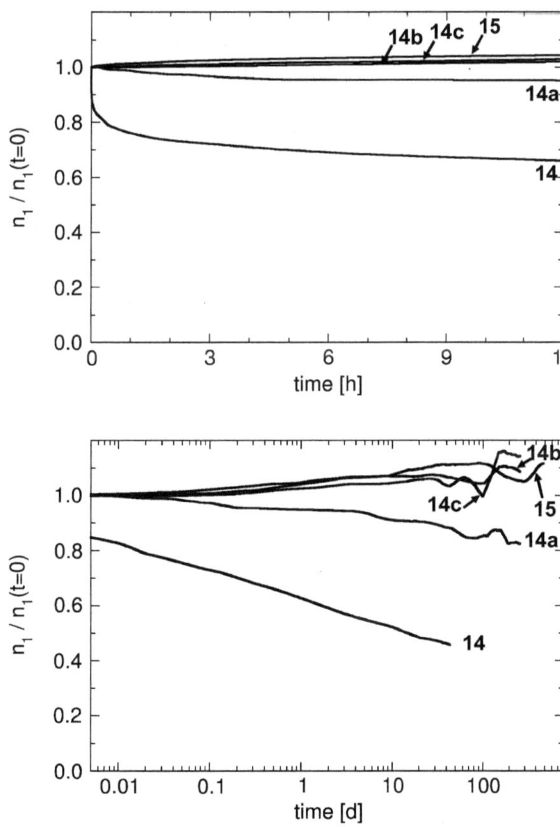

Fig. 31 Long-term stability of holographic gratings for block copolymers **14–14c** and **15**. Note the logarithmic time axis in the lower plot. (Adapted with permission from [72]. Copyright Wiley-VCH Verlag GmbH & Co. KGaA)

14c, and **15**, which contain higher amounts of mesogenic side-groups, the diffraction efficiencies even increased slightly. These gratings are stable for more than two years. From the data it cannot be concluded whether the post-development effect is due to a spontaneous slow increase of the orientation in the illuminated areas of the gratings or a reorientation in the dark regions. Other studies with homogeneous illumination of liquid crystalline azobenzene polymers revealed an increase in orientation after thermal treatment [114]; hence, a similar effect may also occur here in the described block copolymers. More details can be found in [72, 79, 80].

5.2.5 Dynamic Relaxation Processes in Block Copolymers with Non-liquid Crystalline Azobenzene Side-Groups

Whereas block copolymers with liquid crystalline azobenzene moieties show a pronounced stability of the volume gratings and even feature an increase in the refractive index modulation during storage at room temperature, the orientation is much less stable in the case of non-liquid crystalline azobenzene groups. A possible assumption is that the decay of the volume gratings is associated with the glass transition temperature of the material. The verification of this hypothesis is not easy to achieve experimentally in block copolymers with rather short minority blocks because the glass transition temperature cannot easily be measured with common techniques. To elucidate the origin of the decay, both holographic and solid-state NMR experiments were conducted at different temperatures. For the ^{2}H NMR measurements, block copolymers **16a** and **16b** with partially deuterated azobenzene side-groups were synthesized (see Fig. 32) in a three-step synthesis similar to that described in Sect. 2.1 (see Fig. 4) [115]. The covalently linked side-groups differ mainly in the content of the deuterium as well as in its position with respect to the polymer backbone. Both block copolymers with these azobenzene side-groups attached to the shorter block are expected to exhibit non-liquid crystalline phase behavior.

The ^{2}H NMR spectra of the block copolymers **16a** and **16b** were recorded in the temperature range 320–380 K. The polybutadiene minority phase represents less than 7 mol% of the block copolymer and the total amount of the deuterium atoms is only between 0.4 and 0.5 wt%, but solid-state ^{2}H NMR spectra can still be recorded with sufficient resolution. Upon heating, when passing through the glass transition temperature, the polymer chains including the deuterated side-groups become more and more mobile, so that the attached C–^{2}H bonds perform an isotropic, liquid-like reorientation. This leads to a narrow central line typical of a liquid. Since such isotropic motion is only possible if the polymer chain also moves, the appearance of the central line clearly proves that the minority phase undergoes the glass transition. Moreover, the liquid-like spectra are first observed below the glass transition temperature of the solid PS phase ($T_g \approx 370$ K). This indicates that a phase-separated morphology is indeed present with a soft PB and a solid PS phase. In addition, no splitting of the liquid line is observed; thus, no liquid crystalline order is found.

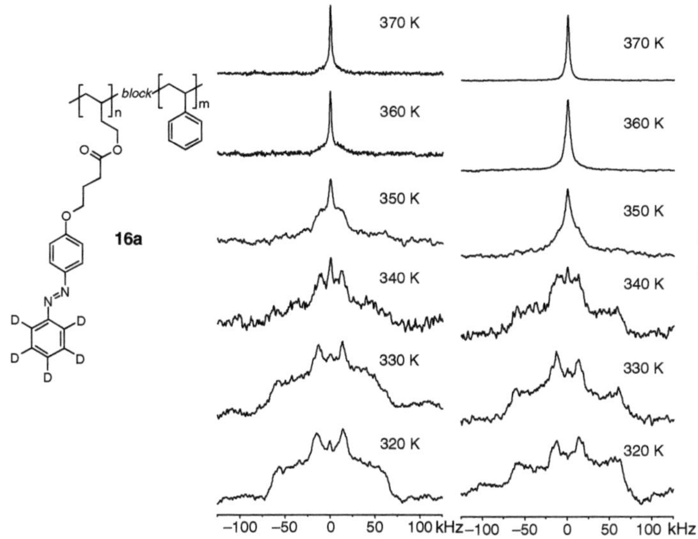

Fig. 32 Schematic representation of block copolymers with partially deuterated azobenzene side-groups **16a** (M_n: 93,000 g mol^{-1}; PDI: 1.04; azobenzene content: 18.3 wt%; ^{2}H-content: 0.54 wt%) and **16b** (M_n: 96,000 g mol^{-1}; PDI: 1.06; azobenzene content: 19.6 wt%; ^{2}H-content: 0.42 wt%) and their temperature-dependent ^{2}H NMR spectra. (Adapted with permission from [115]. Copyright Wiley-VCH Verlag GmbH & Co. KGaA)

From similar investigations it is known that the glass transition temperature T_g lies below the appearance of the liquid line by a factor of approximately 1.25. Therefore, the T_g of the soft PB phase can be determined to 300 K from these ^{2}H NMR data.

It is widely accepted that the decay of the refractive index modulation of holographic gratings consists of several independent processes [75–78] which can be described by the following function:

$$n_1 = (C e^{-k_C t} + D e^{-k_D t} + E) e^{-k_E t}. \tag{8}$$

where C, D, and E are the pre-exponential factors and k_C, k_D, and k_E denote the corresponding rate constants of the different decay processes. The first decay term is mainly attributed to *cis–trans* isomerization, which is significantly accelerated at higher temperatures. The second decay may be attributed to a partial loss of the in-plane orientation due to the free volume present at each given temperature. Finally, the third term describes the decay caused by the movement of the backbone and, thus, probes the glass transition temperature (see Fig. 33). In a plot of the normalized prefactor E_{rel} (E divided by the total refractive index modulation) vs temperature one expects a kink at the glass transition temperature. From the data in Fig. 33 (bottom) one can determine the glass transition temperature to 307 ± 5 K, in good agreement with the value obtained by ^{2}H NMR. These measurements indicate

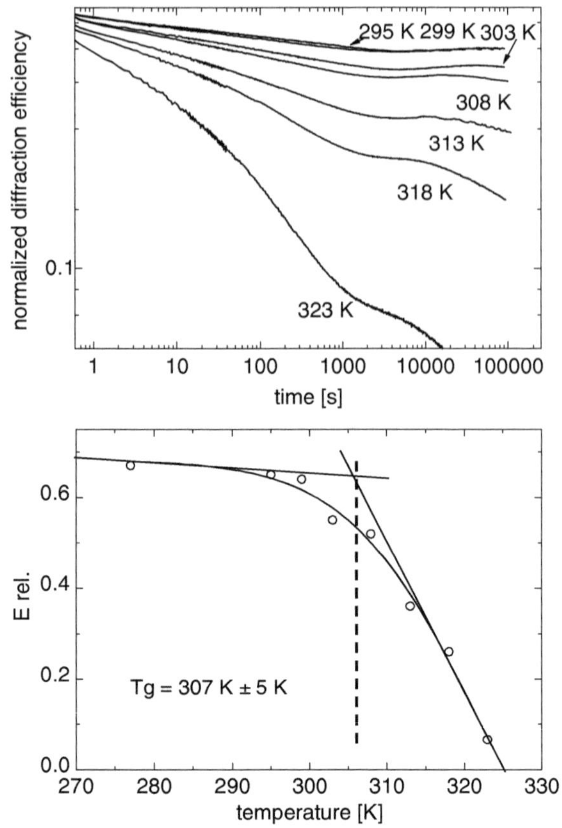

Fig. 33 Dynamic relaxation processes in block copolymer **7** containing the methoxy-substituted azobenzene, as observed by temperature-dependent holographic measurements between 295 and 323 K. (Adapted with permission from [115]. Copyright Wiley-VCH Verlag GmbH & Co. KGaA)

that a fundamental loss of the refractive index modulation in polymers based on a butadiene segment and carrying non-mesogenic azobenzene chromophores is due to the fact that the T_g is close to room temperature, so the backbone is at least partially softened. This is not the case for liquid crystalline systems in which the structural relaxation occurs near the clearing temperature and not near T_g. More details are given in [115].

6 Holographic Data Storage by Angle Multiplexing

A promising way to store an enormous amount of data is to encode the information in holographic gratings. Besides the lateral dimensions x and y of the storage material, the third (depth) dimension z can be employed by several multiplexing techniques. The angular selectivity of thick gratings, for example, permits the independent inscription and reconstruction of holograms at the same lateral spot under different angles θ. The angular resolution and, hence, the multiplexing factor are proportional to the hologram thickness. The presence or absence of a hologram at a

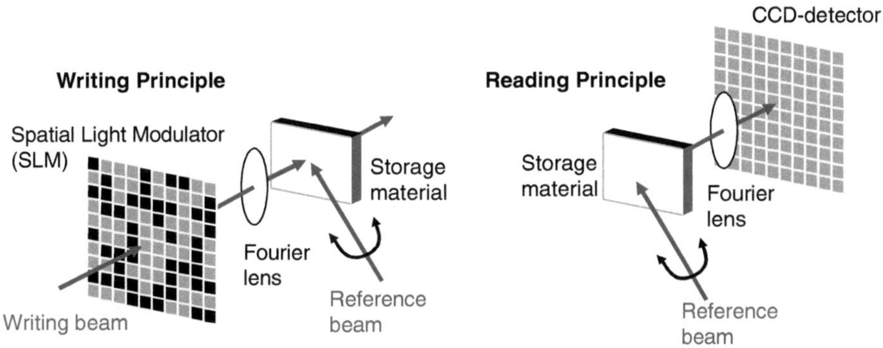

Fig. 34 Parallel holographic inscription of a two-dimensional bit pattern using a spatial light modulator *(left)* and read-out of the two-dimensional bit pattern *(right)*. The reconstructed hologram of the pattern is imaged onto a sensitive CCD detector

specific set of spatial coordinates x, y, and θ encodes a bit of information. For efficient inscription and read-out, it is advisable to write and read entire data pages in parallel. This requires the use of a SLM and a sensitive CCD camera, as is illustrated in Fig. 34.

Materials suitable for angle multiplexing in holographic data storage have to fulfill a number of properties:

- Photosensitivity: a basic condition for holographic data storage is the interaction of the material with light and the formation of refractive index modulation.
- Sample thickness: for angle-selective volume holography, samples of 1–3 mm thickness are required. In this case the Bragg condition ensures that diffraction of the reading beam occurs only if the beam is irradiated at the proper angle. Then it is possible to inscribe and reconstruct many holograms independently at the same lateral position of the material. The thickness of a volume hologram (in the direction of light propagation) must be much larger than its grating spacing (i.e., in the millimeter range).
- Optical density: in order for the light to penetrate the entire thickness of the sample, its optical density at the writing wavelength must not exceed a value of ≈ 0.7.
- Excellent optical quality: a high-density data page with as many as a million pixels encoding digital data must be imaged onto the material and, during reconstruction, its hologram must diffract the reading light toward the detector array, pixel for pixel. For sufficiently low bit error rates, this requires storage materials of very high homogeneity throughout the volume and high-quality optical surfaces, since the lower limit of the diffraction efficiency of holograms is given by spurious light signals due to scattering of the reading beam.
- High dynamic range: with an increasing number of holograms which are recorded in the same volume element, the refractive index modulation and, hence, the diffraction efficiency of each hologram become smaller. The diffraction efficiency scales as the inverse of the square of the number of holograms.

Hence, the maximum number is limited by the signal-to-noise ratio during read-out of the individual holograms (i.e., by the dynamic range of the material).

- High-fidelity recording: especially for the inscription of numerous holograms with low diffraction efficiencies, the material must respond precisely to the light fluence during inscription.
- High sensitivity: to inscribe data at a reasonable rate, the phase gratings of sufficient refractive index modulation must be formed as quickly as possible.
- Shelf-life: for commercial applications, the storage material has to exhibit a sufficient shelf-life prior to the data inscription.
- Long-term stability: the material must retain the inscribed information for long periods of time and it should not be affected by the reading beam.

These requirements are valid for both types of storage media: i.e., write-once and rewritable materials. For write-once, read-many (WORM) storage media, an irreversible material, such as photopolymer systems or dye-doped thermoplastics, can be used, which provides stable gratings after exposure. Irreversible holographic storage materials seem to be fairly advanced, but less is known in the case of reversible storage materials. Our approach for a holographic data storage material is based on azobenzene-containing polymers. Due to the reversible nature of their light-induced reorientations, azobenzene-based storage materials belong to the class of rewritable materials.

As discussed above, angle multiplexing of a large number of holograms requires storage materials with thicknesses in the millimeter range. At the same time, the optical density at the writing wavelength must not be higher than about 0.7. Although materials based on azobenzene are excited by the writing wavelength in the region of the far less intensive n–π^*-transition, the maximum thickness is reached very quickly. Therefore, azobenzene-containing homopolymers cannot be used. Diluting the chromophores by statistical copolymerization causes unstable gratings, since the cooperative effects between the chromophores are lost. Block copolymers which carry the azobenzene units in the minority phase and perform microphase separation are good candidates to reduce the optical density while maintaining the benefits arising from the cooperative effect. However, as shown in Fig. 35, for block copolymers comprising an azobenzene content between 10 and 30 wt%, the limit of the optical density is reached within some tens of microns. In order to achieve samples with a thickness of 1–3 mm and an optical density of 0.5–0.7, the azobenzene content should be in the range of 0.2–0.4 wt%.

Hence, the optical density per micrometer of the azobenzene-containing block copolymers has to be further reduced. One approach is to replace part of the chromophores with mesogenic groups which do not absorb the laser light but can be reoriented together with the azobenzene moieties and enhance the birefringence of the illuminated areas (see Sect. 5.2.4). Another very interesting technical approach to reduce the optical density is to blend the block copolymer with a homopolymer of the same chemical structure as the amorphous photoinactive matrix segment. Blends consisting of at least 95 wt% of a photoinactive homopolymer are required to achieve 1 mm-thick samples with the ideal optical density.

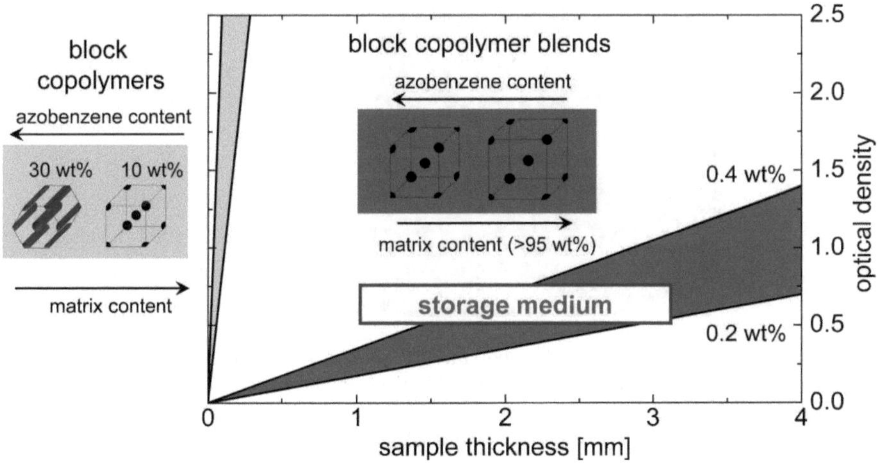

Fig. 35 Schematic representation of the increase in the optical density at the writing wavelength as a function of sample thickness for a neat block copolymer *(light gray)* and a blend material (consisting of the block copolymer with a homopolymer corresponding to the majority block; *dark gray*). The range of 0.2–0.4 wt% of the total azobenzene content in a blend system meets the requirement of a storage medium with a thickness of 1–3 mm. For this – depending on the azobenzene content in the block copolymer – at least 95 wt% of an amorphous, non-absorbing homopolymer is required

6.1 Block Copolymer Blends

In a first step we prepared block copolymer blend samples. Blending with a large amount of the homopolymer will inevitably led to a spherical morphology, but we need to investigate whether the photophysical behavior of the minority segment is maintained or changed.

To do this we have studied blends consisting of an azobenzene-containing block-copolymer based on a PMMA–PHEMA segment and varying amounts of a PMMA homopolymer. As block copolymers, the azobenzene-containing block copolymers **11** (M_n: 42,400 g mol^{-1}; PDI: 1.11; azobenzene content: 25 wt%; see Fig. 28) and **17** (M_n: 152,000 g mol^{-1}; PDI: 1.09; azobenzene content: 24 wt%; see Fig. 36) were used. The block copolymers have similar compositions but different molecular weights. Both block copolymers exhibit a cylindrical morphology in the solid state (see Fig. 36a), but the cylinder diameter varies from 17 nm for **17** to 12 nm for **11**, which is consistent with different bock lengths. Two series of blends were prepared utilizing either block copolymer **11** or **17** together with three different amounts of a PMMA homopolymer (M_n: 40,000 g mol^{-1}; PDI: 1.05). Binary blends were prepared by casting a filtered 5 wt% solution. The solvent were slowly evaporated to allow the formation of the morphology. The polymer films were dried under vacuum followed by annealing at 160°C. All samples showed good optical properties and absence of undesired scattering. TEM images of such blends reveal spheres with a diameter which is very close to the cylinder diameter of the corresponding block

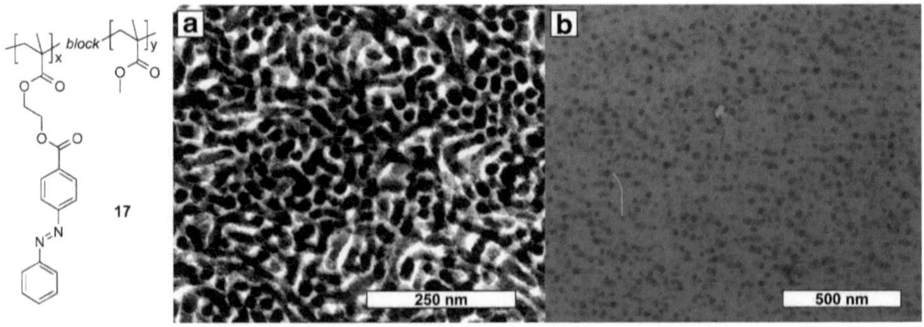

Fig. 36 **a** TEM image of the block copolymer **17** stained with RuO_4 showing a cylindrical morphology (cylinder diameter: 17 ± 3 nm). **b** TEM image of a blend consisting of 36 wt% of block copolymer **11** (cylinder diameter: 12 ± 1.5 nm) and 64 wt% of a PMMA homopolymer yielding a spherical morphology (sphere radius 5.9 ± 0.7 nm). (Adapted with permission from [66]. Copyright 2007 American Chemical Society)

copolymers (see Fig. 36b). The sizes of the spheres for all blends of each series are independent of the added amount of the PMMA homopolymer. This indicates that the homopolymer does not interact with or swell the azobenzene-containing minority phase. Thus, the minority segment is preserved and not affected by the homopolymer addition.

The amount of the homopolymer added to the blend varies between 50 and 84 wt%. Whereas the sphere size remains constant, the distance between the spheres has to increase on average with increasing amount of the homopolymer. SAXS measurements of the different blends confirmed that, with increasing amount of the PMMA homopolymer, the distance increases, thus corroborating the above findings.

As an alternative to holographic experiments, the photoinduced birefringence can also be measured by the response of the material to polarized light. Its value depends on the concentration of the photoactive chromophores, which decreases with increasing amount of the homopolymer. Normalizing the results to the azobenzene content yields very similar results for each series of blends in which the same block copolymer is used (see Fig. 37). This also confirms that the photoresponse is independent of the homopolymer content.

6.2 *Preparation of Thick Samples*

There are several ways to prepare polymer samples with thicknesses in the millimeter range, e.g., cell casting, melt pressing, and injection molding. Injection molding is the most important technology to produce polymer parts and is a well-established industrial process. Therefore, we focused our work on this method. For holographic investigations of thick samples, the block copolymers **18** and **19** were used (see Fig. 38). These block copolymers are based on a PS–PHEMA backbone (ru_{PS}: 528; ru_{HEMA}: 40). Block copolymer **18** consists of 85 wt% polystyrene matrix

Fig. 37 Growth (laser on) and decay (laser off) of the normalized induced birefringence (Δn_{norm}) from polymeric films of binary blends ranging from 68 to 90 μm. *Top*: specimens of blends consisting of azo-functionalized block copolymer **17** and 50 wt% *(solid line)*, 64 wt% *(broken line)* or 84 wt% *(dotted line)* of PMMA homopolymer. *Bottom*: specimens of blends consisting of azo-functionalized block copolymer **11** and 50 wt% *(solid line)*, 71 wt% *(broken line)* or 80 wt% *(dotted line)* of PMMA homopolymer; Δn_{norm} is normalized with respect to the azobenzene content (writing time: 3,600 s, writing wavelength: 532 nm, intensity: 100 mW cm^{-2}, reading wavelength: 633 nm). (Adapted with permission from [66]. Copyright 2007 American Chemical Society)

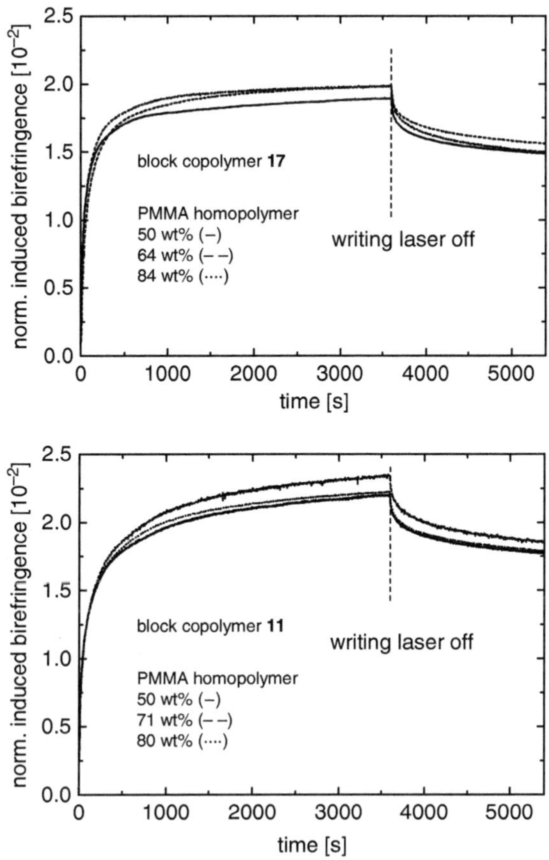

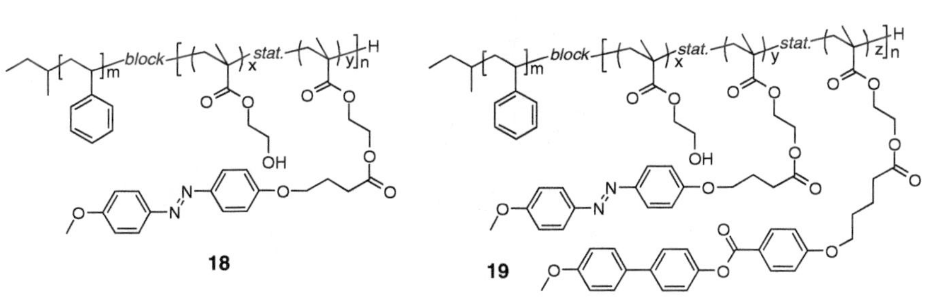

Fig. 38 Schematic representation of the block copolymers used in blends with polystyrene and for the preparation of thick samples by injection molding: Block copolymer **18** (M_n: 65,000 g mol^{-1}; w_{Azo}: 10 wt%; w_{HEMA}: 5 wt%; w_{PS}: 85 wt%. Block copolymer **19** (M_n: 71,000 g mol^{-1}, w_{Azo}: 8.4 wt%, w_{Mesogen}: 12.3 wt%; w_{HEMA}: 1.8 wt%; w_{PS}: 85 wt%. (Reproduced with permission from [67]. Copyright Wiley-VCH Verlag GmbH & Co. KGaA)

segment, 10 wt% of the azobenzene content, and some residual unreacted HEMA repeating units. Block copolymer **19** possesses a 77.5 wt% polystyrene block. In the minority block a random distribution of 8.4 wt% azobenzene side-groups, 12.3 wt% three-ring mesogenic side-groups, and 1.8 wt% hydroxyethyl methacrylate is present. Only 24% of the HEMA moieties carry an azobenzene chromophore. The unreacted HEMA units have no significant influence on the side chain reorientation, but enhance the phase separation.

Binary blends were prepared by mixing a polystyrene homopolymer with the PS–PHEMA block copolymers described above. For the blends, polystyrene with an M_n of 124,000 g mol^{-1} and a PDI of 2, synthesized by a free radical polymerization, was used. In a general solution blending procedure, 20 mL of chloroform were added to 10 g of polymer mixture (5 wt% block copolymer and 95 wt% PS) and slowly rotated in order to achieve a homogeneous solution. The solvent was evaporated off. Air bubbles and remaining solvent were removed by annealing at 180°C under high vacuum. 1.1 mm-thick samples with a diameter of 25 mm were prepared by injection molding. The melt of the polymer blend was injected at a temperature of 200°C with a pressure of 8 bar into a surface-polished mold. After injection molding, the specimen exhibited a pronounced birefringence caused by flow orientation during the injection molding process of the polymer melt. Annealing the polymeric sample inside the mold for 20 h at 130°C eliminated this issue, yielding a homogenous, highly transparent specimen (see Fig. 39 left). The dilution with 95 wt% of PS leads to the required low optical density of about 0.7 at the writing wavelength in a 1 mm-thick sample. These injection-molded samples are suitable for angle multiplexing of holographic gratings. Since the spherical azobenzene-containing domains in these blends have typical sizes of 10–20 nm, the concentration of these spheres in such blends is still between 3,000 and 10,000 spheres per cubic micrometer.

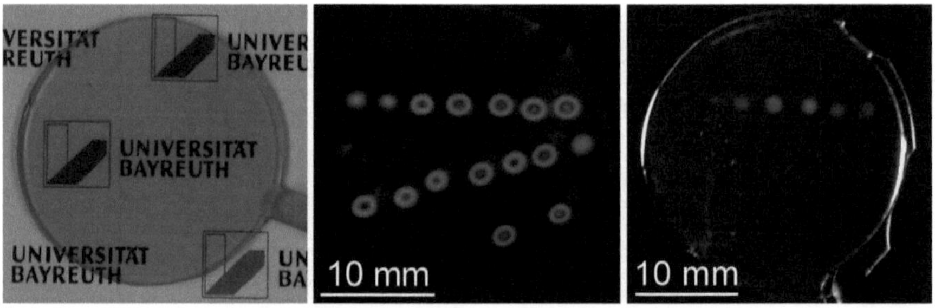

Fig. 39 Photographs of an injection-molded sample (25 mm diameter and 1.1 mm thickness) of polymer blend comprising block copolymer **18**; *left*: injection-molded sample without polarizers; *middle*: sample with several inscribed holograms between crossed polarizers, *right*: sample showing the first diffraction order of some of these holograms under illumination with white light. (Reproduced with permission from [67]. Copyright Wiley-VCH Verlag GmbH & Co. KGaA)

6.3 *Multiplexing of Plane-Wave Holograms*

When several holograms are inscribed at different angles, they must share the maximum achievable refractive index modulation $n_{1,\max}$. Consequently, the refractive index modulation of each hologram is much lower than $n_{1,\max}$.

6.3.1 Intensity vs Polarization Gratings

As a starting point, the effect of subsequent writing of 20 holograms at the same spot was studied in detail for different polarization settings of the laser beams. For the sake of simplicity, plane gratings were used. After each writing process, the diffraction efficiency was measured over the whole angular range covered so far. Figure 40 shows the results of such an experiment when the writing beams had ss- (Fig. 40 top) and sp-polarization (Fig. 40 bottom).

The polarization configuration ss generates a light intensity grating with constant polarization in the material. Figure 40 (top) shows that in this case the last hologram

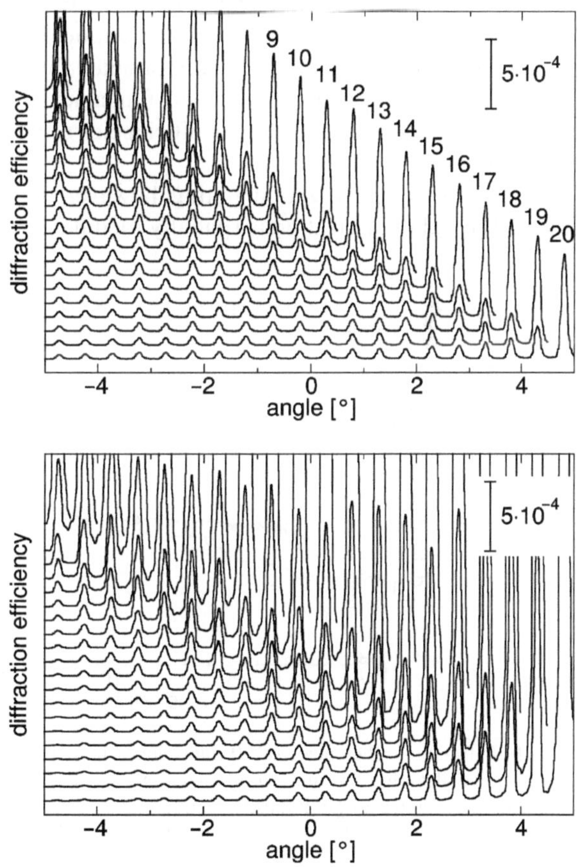

Fig. 40 Angle multiplexing of 20 holographic gratings inscribed in a 1.1 mm-thick injection molded sample of a blend of block copolymer **19** with ss- *(top panel)* and sp-polarization *(bottom panel)*. ss-Polarization corresponds to a light intensity grating, sp-polarization to a polarization grating. Within each panel, from top to bottom the angular dependence of the diffraction efficiency of all holograms is shown after each writing process (vertically offset for clarity). The holograms were inscribed at 514 nm with a fluence of 10 J cm^{-2} of each writing beam; read-out was performed at 685 nm and with s-polarization. (Adapted with permission from [67]. Copyright Wiley-VCH Verlag GmbH & Co. KGaA)

always has a distinctly higher diffraction efficiency than the previous hologram. Its signal is, in turn, sharply reduced by the next writing cycle. At the end of the 20 inscriptions, only the last two gratings show stronger diffraction than the other holograms, all of which have the same diffraction efficiency. The difference of the last two gratings can, at least in part, be ascribed to a non-zero *cis* population of the chromophores in the illuminated areas which enhances the refractive index modulation. Further writing processes eliminate these *cis* population gratings; those of the last writing cycles decay spontaneously with time (see below).

Single polarization gratings lead to a higher refractive index modulation in azobenzene materials than single light intensity gratings. Angle multiplexing with equal exposure, on the other hand, does not result in equal diffraction efficiencies, as Fig. 40 (bottom) shows. Here the polarization setting of the writing beams was sp, corresponding to a polarization grating of constant light intensity. The diffraction efficiencies of the resulting holograms vary in a roughly exponential way: The efficiency of each subsequently written grating is about 20–30% higher than that inscribed immediately before. This is more clearly depicted in Fig. 41.

The difference between intensity and polarization gratings can be understood as follows. If both writing beams are s-polarized, the polarization direction in the material is also s everywhere and is not even altered upon rotation of the sample. Only azobenzenes whose transition dipole moments have a component along the polarization axis can be excited and reoriented. In this configuration, the temporal order of the writing processes has no fundamental influence (except for the last one or two gratings, the diffraction efficiencies of which are not reduced by further inscriptions).

In the case of a polarization grating, however, the situation is fundamentally different. The light polarization is spatially modulated such that chromophores which

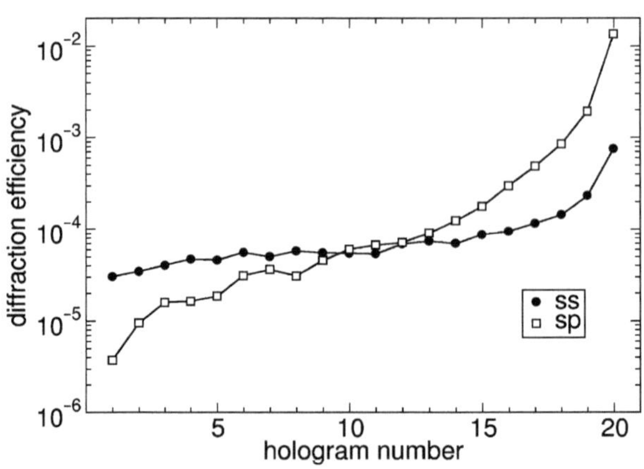

Fig. 41 Diffraction efficiencies of 20 multiplexed holograms written with different polarization settings. The data for ss- and sp-polarization correspond to the final curves in Fig. 40. ss-Polarization generates a light intensity grating and sp-polarization a polarization grating. Reading was performed with s-polarization. (Adapted with permission from [67]. Copyright Wiley-VCH Verlag GmbH & Co. KGaA)

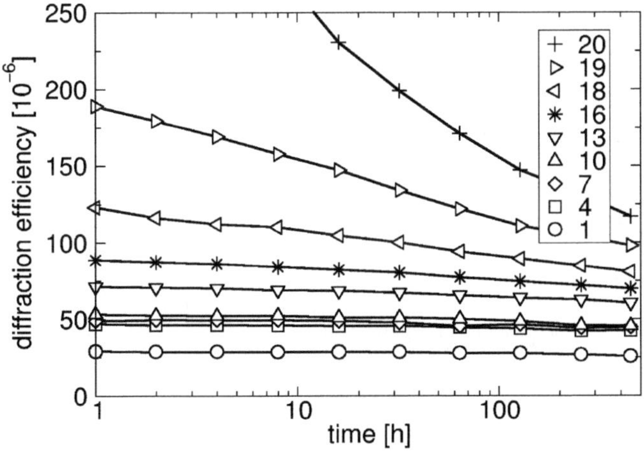

Fig. 42 Temporal decay of some of the 20 holograms (see Fig. 40 top) during a period of 3 weeks. (Adapted with permission from [67]. Copyright Wiley-VCH Verlag GmbH & Co. KGaA)

have been oriented in a previous inscription process can be reoriented in different directions in subsequent cycles. Hence, later writing processes will efficiently erase earlier gratings. In order to achieve roughly uniform diffraction efficiencies for all holograms, the exposure must be widely varied over the sequence of inscriptions.

For holographic information storage, both types of gratings have advantages and disadvantages. Intensity gratings permit multiplexing with more or less equal exposure times. Polarization gratings applied to azobenzene polymers, on the other hand, have a unique feature not present in most other holographic storage materials. They reconstruct not only the intensity and phase of the original object beam but also its polarization [116]. The reconstructed polarization can even be perpendicular to the polarization of the reference beam. This makes it largely possible to eliminate light-scattering problems, since the polarizations of scattered light and reconstructed image are then perpendicular to each other [21].

The stability of the 20 grating written with ss-polarization was recorded over a period of 3 weeks. The decay of the diffraction efficiencies is plotted in Fig. 42.

The efficiencies of the first 15 gratings are in the same range and very stable during the whole 3-week period. The last five holograms show decay, yet their signals still remain above those of the earlier written gratings. Part of this behavior can be ascribed to the relaxation of the *cis* isomers to the *trans* state; another contributing effect may be an increased amount of free volume around some of the chromophores after the light-induced reorientation.

6.3.2 Possibilities for Angle Multiplexing

It is possible to inscribe many more multiplexed holograms at the same position with these azobenzene-containing polymer blend samples. The limit is determined by the

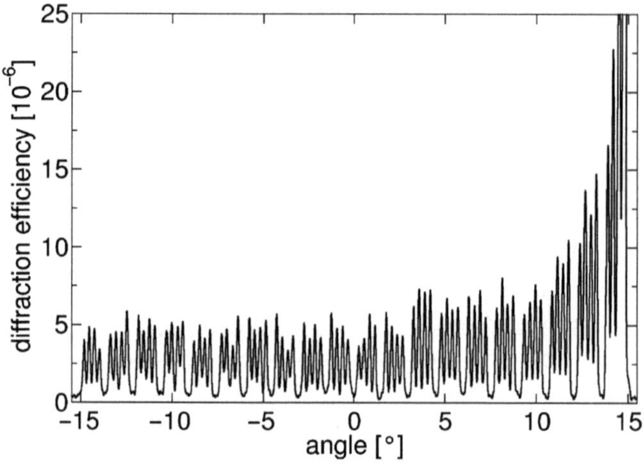

Fig. 43 Eighty angle-multiplexed holograms each written with a fluence of $5\,\mathrm{J\,cm^{-2}}$. Sequences of four holograms were inscribed and every fifth was omitted. (Adapted with permission from [67]. Copyright Wiley-VCH Verlag GmbH & Co. KGaA)

signal-to-noise ratio, since all holograms must share the maximum refractive index modulation provided by the material. The diffraction efficiency of 80 multiplexed gratings is shown in Fig. 43 as a function of sample orientation. Here a sequence of four holograms was inscribed and every fifth was omitted.

Even for this number, roughly uniform diffraction efficiencies were achieved by writing all of them with the same exposure. The current limiting number of angle-multiplexed gratings for which clear discrimination levels of the diffraction efficiency for the presence (logical "1") and absence (logical "0") of a grating can be defined is around 200 (data not shown). It should be possible to increase this number if the amount of light scattering of the material (and, hence, the detected background signal during reading) can be further reduced.

6.4 Multiple Read/Write Cycles

Since the reorientation of the azobenzene side chains is reversible, one can erase all information written in the entire medium by heating it above the glass transition temperature. However, a purely optical method which also permits local erasure would be preferable. It has been reported that in thin azobenzene-containing films, many successful rewriting cycles are possible when deleting is performed with UV light [117, 118]. This method is not applicable to thick samples in the millimeter range, however, since the high absorption coefficient of the polymer in the UV region prevents penetration through the whole material. Deleting with a single beam of the writing wavelength has also been reported [15, 119, 120]. In those studies, the deleted areas were not rewritten. Here, a different method is presented which allows a very large number (more than 1,000) of successful writing and erasure cycles to be performed.

The method consists of two steps. The first is the inscription of a holographic grating which is phase-shifted by 180° with respect to the original grating to be erased. In a second step, the chromophores are oriented back by a single laser beam that is polarized perpendicularly to the writing beams. If a 2D bit pattern was written, the first step would consist of inscribing the same information again with the reference beam phase-shifted by 180°.

Figure 44 demonstrates as many as 1,000 writing and deleting cycles of a plane holographic grating.

During the first few rewriting cycles the diffraction efficiency even increases. This effect indicates an increasing degree of orientation of the chromophores in the polarization direction of the writing beams. After about 10 cycles, the diffraction efficiency reaches a stable value with no significant degradation, even after 1,000 cycles. In the erasure processes the diffraction efficiency is reduced by more than two orders of magnitude.

To our knowledge, this is the first time that long-term stability of angle-multiplexed volume holographic gratings has been achieved in a rewritable polymeric medium, in which writing, reading, and deleting is possible by purely optical means.

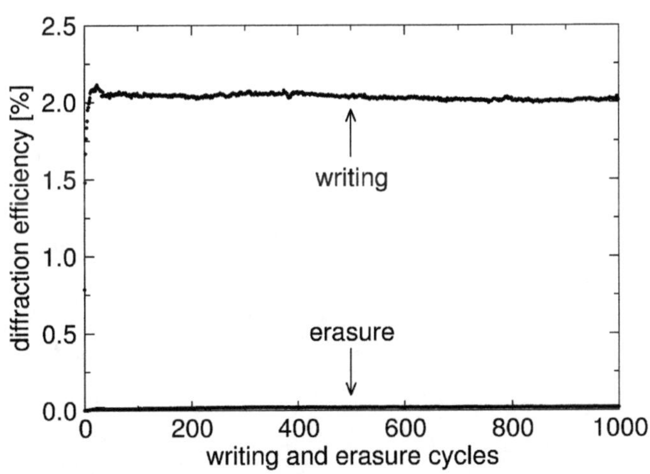

Fig. 44 Sequence of 1,000 writing and erasure cycles of volume gratings in samples of a polymer blend of block copolymer **19**. In each cycle, an s-polarized plane holographic grating was written with a fluence of 10 $J cm^{-2}$ at 514 nm. Deleting was performed by inscribing a grating phase-shifted by 180° with the same fluence and subsequent irradiation of a single p-polarized beam of fluence 50 $J cm^{-2}$. (Adapted with permission from [67]. Copyright Wiley-VCH Verlag GmbH & Co. KGaA)

6.5 Parallel Inscription of Data Pages Using a Spatial Light Modulator

For technical applications of volume holography in the field of data storage, parallel inscription and read-out of angle-multiplexed data pages are highly desirable for achieving sufficiently fast data rates. This can be accomplished with a SLM, which generates the pixel pattern during inscription, and a CCD camera for reading, as shown in Fig. 34. Initial experimental results are presented in Fig. 45.

Eight patterns of arbitrary characters and numbers were generated and holographically stored in an injection molded specimen of a polymer blend comprising 10.9 wt% of block copolymer **18** (Fig. 38 left) and a polystyrene homopolymer with a number average molecular weight of 124,000 g mol^{-1}. Between each two writing processes the sample was rotated by 1°. The reconstructed patterns are clearly recognizable, and although the quality and the sharpness are not sufficient, the principal function is demonstrated.

Basically two problems were encountered. The first is related to the storage process. Since the Fourier transforms of the patterns are inscribed, the light intensity distribution in the material represents their spatial frequencies. At the center of all holograms, a small bright spot is present which corresponds to low frequencies $k_x \approx k_y \approx 0$. The corresponding high fluences of all holograms add up in the central region, so a detailed reconstruction requires a high dynamic range of the storage material. This requirement can, in principle, be avoided if an SLM is used which allows for an adjustment of the phase shift of the transmitted light. In this case, the light phase can be shifted by 180° between subsequent inscriptions and the effective exposure in the central region remains small [121]. With the light modulator used here, such a phase adjustment was not possible.

The second issue is scattered light which leads to the tiny bright dots in the reconstructed patterns and is far more detrimental than in the case of plane-wave

Fig. 45 Reconstruction of eight two-dimensional images which were inscribed at the same spot of a thick sample of material **18** under angle intervals of 1° each. The object beam had a diameter of about 1 mm in the sample. Writing was performed with a wavelength of 514 nm, ss-polarization, and a total fluence of 2 J cm^{-2} per hologram. (Reprinted with permission from [79]. Copyright 2009 SPIE)

holograms. Scattering is caused by inhomogeneities in the bulk and/or the surface of the specimen. Improved preparation techniques are expected to reduce the amount of inhomogeneities and, hence, light scattering. An alternative way of eliminating it may be to inscribe polarization gratings rather than intensity gratings, because in this case the diffracted light of the reconstruction has the same polarization as the object beam during inscription, which can be chosen perpendicular to the polarization direction of the reading beam. The scattered light is then also polarized perpendicularly to the reconstructed data and can be efficiently blocked. A disadvantage of polarization gratings is that they must be inscribed with strongly different exposures, as was discussed in Sect. 6.3.1.

7 Summary and Conclusions

We have demonstrated that azobenzene-containing materials can efficiently form holographic SRGs and volume gratings for various kinds of applications. Two material classes, azobenzene-containing block copolymers and molecular glasses, were designed and synthesized and they can find applications in different fields.

SRGs are detrimental to the angle-multiplexed inscription of volume holograms and must be avoided for high-density data storage applications. SRGs can be efficiently suppressed in block copolymers carrying the azobenzene chromophores as side chains in their minority block. Regarding the refractive index modulation and the stability of volume gratings, block copolymers are far superior to statistical copolymers. This shows that the cooperative effects between the azobenzene chromophores in the minority segments of block copolymers are largely preserved, which improves the absolute value as well as the stability of the refractive index modulation. Using appropriate chromophores, e.g., liquid-crystalline cyano-azobenzenes, yields a further improvement. With azobenzene side-groups not possessing liquid-crystalline properties, a similar stabilization is observed if they are combined with mesogenic side-groups.

Blending the block copolymers with different amounts of the homopolymer corresponding to the majority phase of the block copolymer does not affect the photophysics of the azobenzene chromophores. In this way the optical density per micrometer at the writing wavelength can be adjusted to the desired value and thick samples can be prepared by conventional injection molding. With an optical density of 0.7 they contain about 3,000–10,000 functionalized spheres per cubic micrometer. These materials are suitable for angle-multiplexing. We have demonstrated so far that 200 holograms can be superimposed and reconstructed independently at the same spot and more than 1,000 write/erase cycles can be performed without significant degradation of the diffraction efficiency. Hence, a rewritable polymeric medium has been presented for the first time, in which writing, reading, and deleting is possible by purely optical means and in which the gratings are stable over at least 2 years. Two-dimensional images were inscribed and read out in parallel with an SLM and a sensitive CCD camera. Future improvements will focus on faster writing speeds and lower light scattering of the material, so the reconstruction, especially of entire data pages, can be performed with a better signal-to-noise ratio.

Azobenzene-containing molecular glasses are a promising class for the controlled formation of surface relief structures. They exhibit superior performance as compared to azobenzene-containing polymers in terms of writing times and modulation heights. SRGs of more than 600 nm can be achieved within a few minutes of illumination. The light-driven material transport on a macroscopic length scale takes place far below the glass transition temperature. The process depends strongly on the polarization state of the writing beams and the electrical susceptibility of the material as described by the gradient force model by Kumar et al. The surface patterns are stable enough to be transferred to a polymer surface via replica molding techniques. Possible applications include HOEs which can be applied as security features.

Acknowledgments The authors are deeply indebted to Dr. Thomas Breiner, Dr. Carsten Frenz, Dr. Michael Häckel, and Dr. Ulrich Theissen for their contributions and their dedicated work during their Ph.D. theses at the University of Bayreuth. Daniela Kropp and Christina Löffler (Makromolekulare Chemie I) are gratefully acknowledged for their invaluable contributions in material synthesis and sample preparation.

HWS wishes to express his special gratitude to Helmut Ringsdorf for inspiring him in 1982 to synthesize and study the first azobenzene side chain liquid-crystalline polymers in his diploma thesis. As this chapter clearly demonstrates, the topic has now broad application potentials and is still fascinating to us.

The authors are grateful to the German Science Foundation for generously providing financial support for this work within the framework of the collaborative research centre SFB 481, project B2.

References

1. Ashley J, Bernal MP, Burr GW, Coufal H, Guenther H, Hoffnagle JA, Jefferson CM, Marcus B, Macfarlane RM, Shelby RM, Sincerbox GT (2000) IBM J Res Dev 44:341–368
2. Hesselink L, Orlov SS, Bashaw MC (2004) Proc IEEE 92:1231–1280
3. Schilling ML, Colvin VL, Dhar L, Harris AL, Schilling FC, Katz HE, Wysocki T, Hale A, Blyler LL, Boyd C (1999) Chem Mater 11:247–254
4. Barbastathis G, Psaltis D (2000) Volume holographic multiplexing methods. In: Coufal H, Psaltis D, Sincerbox GT (eds) Holographic data storage. Springer, Berlin
5. Waldmann DA, Butler CJ, Raguin DH (2003) Proc SPIE 5216:10–25
6. Schnoes M, Ihas B, Dhar L, Michaels D, Settachayanon S, Schomberger GL, Wilson WL (2003) Proc SPIE 4988:68–76
7. Zhao Y, Ikeda T (2009) Smart light-responsive materials. Wiley, Hoboken, NJ
8. Hagen R, Bieringer T (2001) Adv Mater 13:1805–1810
9. Ringsdorf H, Schmidt H-W (1984) Makromol Chem 185:1327–1334
10. Eich M, Wendorff JH, Ringsdorf H, Schmidt H-W (1985) Macromol Chem Phys 186: 2639–2647
11. Eich M, Wendorff JH, Reck B, Ringsdorf H (1987) Macromol Rapid Commun 8:59–63
12. Meng X, Natansohn A, Rochon P (1997) Polymer 38:2677–2682
13. Cimrova V, Neher D, Kostromine S, Bieringer T (1999) Macromolecules 32:8496–8503
14. Hvilsted S, Andruzzi F, Kullia C, Siesler HW, Ramanujam PS (1995) Macromolecules 28:2172–2183

15. Berg RH, Hvilsted S, Ramanujam PS (1996) Nature 383:505–508
16. Ringsdorf H, Schmidt HW, Baur G, Kiefer R, Windscheid F (1986) Liq Cryst 1:319–325
17. Zilker SJ, Bieringer T, Haarer D, Stein RS, Van Egmond JW, Kostromine SG (1998) Adv Mater 10:855–859
18. Zilker SJ, Huber MR, Bieringer T, Haarer D (1999) Appl Phys B 68:893–897
19. Rochon P, Batalla E, Natansohn A (1995) Appl Phys Lett 66:136–138
20. Kim DY, Tripathy SK, Li L, Kumar J (1995) Appl Phys Lett 66:1166–1168
21. Minabe J, Maruyama T, Yasuda S, Kawano K, Hayashi K, Ogasawara Y (2004) Jpn J Appl Phys 43:4964–4967
22. Tian Y, Watanabe K, Kong X, Abe J, Iyoda T (2002) Macromolecules 35:3739–3747
23. Cui L, Zhao Y, Yavrian A, Galstian T (2003) Macromolecules 36:8246–8252
24. Cui L, Tong X, Yan X, Liu G, Zhao Y (2004) Macromolecules 37:7097–7104
25. Han Y-K, Dufour B, Wu W, Kowalewski T, Matyjaszewski K (2004) Macromolecules 37:9355–9365
26. Wang G, Tong X, Zhao Y (2004) Macromolecules 37:8911–8917
27. Ravi P, Sin SL, Gan LH, Gan YY, Tam KC, Xia XL, Hu X (2005) Polymer 46:137–146
28. Forcen P, Oriol L, Sanchez C, Alcala R, Hvilsted S, Jankova K, Loos J (2007) J Polym Sci Polym Chem 45:1899–1910
29. Forcen P, Oriol L, Sanchez C, Rodriguez FJ, Alcala R, Hvilsted S, Jankova K (2007) Eur Polym J 43:3292–3300
30. Tong X, Cui L, Zhao Y (2004) Macromolecules 37:3101–3112
31. Sin SL, Gan LH, Hu X, Tam KC, Gan YY (2005) Macromolecules 38:3943–3948
32. Yu H, Shishido A, Ikeda T, Iyoda T (2005) Macromol Rapid Commun 26:1594–1598
33. Kadota S, Aoki K, Nagano S, Seki T (2006) Colloids Surf 284/285:535–541
34. Morikawa Y, Kondo T, Nagano S, Seki T (2007) Chem Mater 19:1540–1542
35. Tang X, Gao L, Fan X, Zhou Q (2007) J Polym Sci A 45:2225–2234
36. Tang X, Gao L, Fan X, Zhou Q (2007) J Polym Sci A 45:5190–5198
37. Zhang Y, Zhang W, Chen X, Cheng Z, Wu J, Zhu J, Zhu X (2008) J Polym Sci A 46:777–789
38. Ding L, Mao H, Xu J, He J, Ding X, Russell TP, Robello DR, Mis M (2008) Macromolecules 41:1897–1900
39. He X, Sun W, Yan D, Xie M, Zhang Y (2008) J Polym Sci A 46:4442–4450
40. Yu H, Naka Y, Shishido A, Ikeda T (2008) Macromolecules 41:7959–7966
41. Gimeno S, Forcen P, Oriol L, Pinol M, Sanchez C, Rodriguez FJ, Alcala R, Jankova K, Hvilsted S (2009) Eur Polym J 45:262–271
42. Yoshida E, Ohta M (2005) Colloid Polym Sci 283:521–531
43. Mao G, Wang J, Clingman SR, Ober CK (1997) Macromolecules 30:2556–2567
44. Osuji CO, Chen JT, Mao G, Ober CK, Thomas EL (2000) Polymer 41:8897–8907
45. Hayakawa T, Horiuchi S, Shimizu H, Kawazoe T, Ohtsu M (2002) J Polym Sci Polym Chem 40:2406–2414
46. Shirota Y (2000) J Mater Chem 10:1–25
47. Strohriegl P, Grazulevicius JV (2002) Adv Mater 14:1439–1452
48. Van der Auweraer M, De Schryver FC, Borsenberger PM, Bassler H (1994) Adv Mater 6:199–213
49. Tang CW, VanSlyke SA (1987) Appl Phys Lett 51:913–915
50. Thelakkat M, Schmidt HW (1998) Adv Mater 10:219–223
51. Bach U, Lupo D, Comte P, Moser JE, Weissortel F, Salbeck J, Spreitzer H, Gratzel M (1998) Nature 395:583–585
52. Thelakkat M, Schmitz C, Hohle C, Strohriegl P, Schmidt HW, Hofmann U, Schloter S, Haarer D (1999) Phys Chem Chem Phys 1:1693–1698
53. Schmitz C, Posch P, Thelakkat M, Schmidt HW (1999) Phys Chem Chem Phys 1:1777–1781
54. Shirota Y (2005) J Mater Chem 15:75–93
55. Kim M-J, Seo E-M, Vak D, Kim D-Y (2003) Chem Mater 15:4021–4027
56. Chun C, Kim M-J, Vak D, Kim DY (2003) J Mater Chem 13:2904–2909
57. Ishow E, Lebon B, He Y, Wang X, Bouteiller L, Galmiche L, Nakatani K (2006) Chem Mater 18:1261–1267

58. Stracke A, Wendorff JH, Goldmann D, Janietz D (2000) Liq Cryst 27:1049–1057
59. Kulikovska O, Goldenberg LM, Kulikovsky L, Stumpe J (2008) Chem Mater 20:3528–3534
60. Adams J, Gronski W (1989) Macromol Rapid Commun 10:553–557
61. Frenz C, Fuchs A, Schmidt H-W, Theissen U, Haarer D (2004) Macromol Chem Phys 205:1246–1258
62. Bates FS, Rosedale JH, Bair HE, Russell TP (1989) Macromolecules 22:2557–2564
63. Sänger J, Tefehne C, Lay R, Gronski W (1996) Polym Bull 36:19–26
64. Frenz C (2003) Diblock copolymers with photoaddressable chromophores for holographic data storage. PhD Thesis, University of Bayreuth
65. Stevens H, Rehage G, Finkelmann H (1984) Macromolecules 17:851–856
66. Breiner T, Kreger K, Hagen R, Häckel M, Kador L, Mueller AHE, Kramer EJ, Schmidt H-W (2007) Macromolecules 40:2100–2108
67. Häckel M, Kador L, Kropp D, Schmidt H-W (2007) Adv Mater 19:227–231
68. Shirota Y, Moriwaki K, Yoshikawa S, Ujike T, Nakano H (1998) J Mater Chem 8:2579–2581
69. Fuhrmann T, Tsutsui T (1999) Chem Mater 11:2226–2232
70. Nakano H, Takahashi T, Kadota T, Shirota Y (2002) Adv Mater 14:1157–1160
71. Audorff H, Walker R, Kador L, Schmidt HW (2009) J Phys Chem B 113:3379–3384
72. Häckel M, Kador L, Kropp D, Frenz C, Schmidt H-W (2005) Adv Funct Mater 15:1722–1727
73. Kogelnik H (1969) Bell Syst Tech J 48:2909–2947
74. Magnusson R, Gaylord TK (1978) J Opt Soc Am 68:806–809
75. Natansohn A, Rochon P (2002) Chem Rev 102:4139–4175
76. Song OK, Wang CH, Pauley MA (1997) Macromolecules 30:6913–6919
77. Boehm N, Materny A, Kiefer W, Steins H, Mueller MM, Schottner G (1996) Macromolecules 29:2599–2604
78. Brown D, Natansohn A, Rochon P (1995) Macromolecules 28:6116–6123
79. Häckel M, Kador L, Kropp D, Frenz C, Schmidt H-W (2005) Proc SPIE 5939:593908/1–593908/10
80. Häckel M, Kador L, Frenz C, Schmidt H-W (2004) Proc SPIE 5521:63–72
81. Carvalho LL, Borges TFC, Cardoso MR, Mendonca CR, Balogh DT (2006) Eur Polym J 42:2589–2595
82. Schulz BM, Huber MR, Bieringer T, Krausch G, Zilker SJ (2001) Synth Met 124:155–157
83. Yager KG, Tanchak OM, Godbout C, Fritzsche H, Barrett CJ (2006) Macromolecules 39:9311–9319
84. Viswanathan NK, Kim DY, Bian S, Williams J, Liu W, Li L, Samuelson L, Kumar J, Tripathy SK (1999) J Mater Chem 9:1941–1955
85. Rochon P, Natansohn A, Callender CL, Robitaille L (1997) Appl Phys Lett 71:1008–1010
86. Delaire JA, Nakatani K (2000) Chem Rev 100:1817–1845
87. Watanabe O, Tsuchimori M, Okada A, Ito H (1997) Appl Phys Lett 71:750–752
88. Kato J, Yamaguchi I, Tanaka H (1996) Opt Lett 21:767–769
89. Pedersen TG, Johansen PM (1997) Phys Rev Lett 79:2470–2473
90. Pedersen TG, Johansen PM, Holme NCR, Ramanujam PS, Hvilsted S (1998) Phys Rev Lett 80:89–92
91. Barrett CJ, Natansohn AL, Rochon PL (1996) J Phys Chem 100:8836–8842
92. Barrett CJ, Rochon PL, Natansohn AL (1998) J Chem Phys 109:1505–1516
93. Yager KG, Barrett CJ (2006) Macromolecules 39:9320–9326
94. Henneberg O, Geue T, Saphiannikova M, Pietsch U, Rochon P, Natansohn A (2001) Appl Surf Sci 182:272–279
95. Leffin P, Fiorini C, Nunzi J-M (1998) Opt Mater 9:323–328
96. Viswanathan NK, Balasubramanian S, Li L, Kumar J, Tripathy SK (1998) J Phys Chem B 102:6064–6070
97. Percec V, Schlueter D, Kwon YK, Blackwell J, Moeller M, Slangen PJ (1995) Macromolecules 28:8807–8818
98. Xiang M, Li X, Ober CK, Char K, Genzer J, Sivaniah E, Kramer EJ, Fischer DA (2000) Macromolecules 33:6106–6119
99. Wang J, Mao G, Ober CK, Kramer EJ (1997) Macromolecules 30:1906–1914

100. You F, Paik MY, Häckel M, Kador L, Kropp D, Schmidt H-W, Ober CK (2006) Adv Funct Mater 16:1577–1581
101. Paik MY, Krishnan S, You F, Li X, Hexemer A, Ando Y, Kang SH, Fischer DA, Kramer EJ, Ober CK (2007) Langmuir 23:5110–5119
102. Ando H, Takahashi T, Nakano H, Shirota Y (2003) Chem Lett 32:710–711
103. Chun C, Ghim J, Kim M-J, Kim DY (2005) J Polym Sci Polym Chem 43:3525–3532
104. Nakano H, Takahashi T, Tanino T, Shirota Y (2007) J Photopolym Sci Technol 20:87–89
105. Yang K, Yang S, Kumar J (2006) Phys Rev B 73:165204–1–14
106. Lefin P, Fiorini C, Nunzi J-M (1998) Opt Mater 9:323–328
107. Kumar J, Li L, Jiang XL, Kim DY, Lee TS, Tripathy S (1998) Appl Phys Lett 72:2096–2098
108. Seo E-M, Kim MJ, Shin Y-D, Lee J-S, Kim D-Y (2001) Mol Cryst Liquid Cryst 370:143–146
109. Reinke N, Draude A, Fuhrmann T, Franke H, Lessard RA (2004) Appl Phys B 78:205–209
110. Sobolewska A, Miniewicz A (2007) J Phys Chem B 111:1536–1544
111. Walker R, Audorff H, Kador L, Schmidt HW (2009) Adv Funct Mater 19:2630–2638
112. Natansohn A, Rochon P, Gosselin J, Xie S (1992) Macromolecules 25:2268–2273
113. Breiner T (2001) Block coplymers with functionalized poly(2-hydroxyethyl methacrylate) segments for optical data storage. PhD Thesis, University of Bayreuth
114. Kidowaki M, Jujiwara T, Morino S, Ichimura K, Stumpe J (2000) Appl Phys Lett 76:1377–1379
115. Kreger K, Loeffler C, Walker R, Wirth N, Bingemann D, Audorff H, Roessler EA, Kador L, Schmidt H-W (2007) Macromol Chem Phys 208:1530–1541
116. Kawano K, Ishii T, Minabe J, Niitsu T, Nishikata Y, Baba K (1999) Opt Lett 24:1269–1271
117. Chen AG, Brady DJ (1993) Appl Phys Lett 62:2920–2922
118. Holme NCR, Ramanujam PS, Hvilsted S (1996) Opt Lett 21:902–904
119. Jiang XL, Li L, Kumar J, Kim DY, Tripathy SK (1998) Appl Phys Lett 72:2502–2504
120. Lagugné-Labarthet F, Buffeteau T, Sourisseau C (2002) Phys Chem Chem Phys 4:4020–4029
121. Jang J-S, Shin D-H (2001) Opt Lett 26:1797–1799

Adv Polym Sci (2010) 228: 123–153
DOI:10.1007/12_2009_34

Published online: 12 February 2010

Donor–Acceptor Block Copolymers with Nanoscale Morphology for Photovoltaic Applications

Michael Sommer, Sven Huettner, and Mukundan Thelakkat

Abstract Extensive research activities in synthesis and device engineering have been devoted to the development of donor–acceptor (D–A) bulk heterojunction solar cells in the last few years. Several photophysical processes occur in such devices which have to be optimized for an efficient device operation. First, excitons that are created upon light absorption need to reach the D/A interface within their short exciton diffusion length (10–20 nm), where they may dissociate into holes and electrons. Subsequent charge transport and charge collection can then take place at the electrodes, given that co-continuous pathways of donor and acceptor domains are provided. An active layer thickness of 100–200 nm is required to absorb most of the light, and vertically aligned pathways with a high aspect ratio of either phase should percolate through the film, taking into account the small exciton diffusion lengths. The morphologies resulting from this ideal situation resemble those of vertically oriented microphase separated block copolymer thin films, and hence suggest the exploitation of D–A block copolymers for organic photovoltaics. Furthermore, complex block copolymer architectures are not only desired in order to improve the morphological control but also to enhance the long term stability of the device. The potential of such block copolymers to microphase separate into well-defined structures several tens of nanometers in size thus addresses the morphological requirements mentioned above. This chapter gives an overview of the emerging field of D–A block copolymers. General synthetic efforts that have been undertaken towards this direction are summarized. The D–A block copolymers prepared in our group are reviewed and complemented with recent work on crystalline–crystalline block copolymers.

M. Sommer, S. Huettner, and M. Thelakkat (✉)
Applied Functional Polymers, Department of Macromolecular Chemistry I,
University of Bayreuth, Universitätsstrasse 30, 95440 Bayreuth, Germany
e-mail: mukundan.thelakkat@uni-bayreuth.de

S. Huettner
Cavendish Laboratory, Thin Films and Interfaces Group – BSS Physics, JJ Thomson Avenue,
Cambridge CB3 0HE, UK

Keywords Donor-acceptor block copolymers · Grignard metathesis polymerization · Morphology · Nitroxide mediated radical polymerization · Organic photovoltaics · Perylene bisimide · Poly(3-hexylthiophene) · Solar cells · Side chain crystallinity

Contents

1 Introduction

Since the first report of an efficient organic photovoltaic (OPV) device made by C. Tang et al. in 1986 [1], power conversion efficiencies (PCEs) are evolving together with an increased understanding of the underlying fundamental processes occurring in active layers of an electron donor (D) and an electron acceptor (A) [2]. The ongoing enhancement in device performance can partially be attributed to the development of novel materials with favourable electronic properties [3–5]. More importantly, intense efforts concerning the optimization of the active layer blend morphology have substantially contributed to the development of PCEs that now range between 4% and 6% [6–10]. Since charge generation and charge recombination in organic bulk heterojunction solar cells occur at the D–A interface, precise control of the size and the shape of this interface is crucial in order to achieve further improvements. From a materials chemists' point of view it is therefore essential to develop new materials as well as new concepts that allow to improve morphology control.

A higher level of morphological control can be achieved by exploiting various interactions between either the same molecules of donor or acceptor, or between the donor and the acceptor molecule. Furthermore, asymmetric surface fields of the device can induce a vertically segregated morphology suitable for bulk heterojunction solar cells [11]. For example, processing a blend of poly(3-hexylthiophene) (P3HT) and phenyl-C_{61}-butyric acid methyl ester (PCBM) blend together with a small amount of a fluorinated PCBM derivative led to the formation of a monolayer with the fluorinated side groups oriented to the film–air interface [12]. With the aim of creating supramolecular pathways of donor and acceptor groups, π–π interactions

[13] and hydrogen bonds [14] were exploited. In another study, hydrogen bonds between only the acceptor moiety perylene bisimide (PBI) were introduced as a structure-directing tool [15]. As a result, thin films comprised of a three-dimensional mesoscopic acceptor network could be fabricated. Most strikingly, this acceptor network was maintained even when processing the material together with an amorphous donor polymer, giving a D–A heterojunction with defined morphology, charge transport pathways and domain sizes in the range of the exciton diffusion length [16]. This result is encouraging since PBIs tend to form very large one-dimensional stacks and crystals [17, 18], which has restricted the use of this type of acceptor in organic solar cells up to now [19–21]. Even stronger interactions for interface tuning can be provided by a covalent bond between the donor and the acceptor moiety. Manifold architectures have been designed, e.g. D–A dyads [22] or double-cable polymers [23, 24], just to name a few. The motivation for the often challenging synthesis was clearly driven by the advantage of a highly intermixed D–A morphology capable of efficient charge separation. Once the charges are separated, they have to percolate towards the respective electrodes. This process occurs most efficiently when the transport pathways between the two electrodes are straight and do not exhibit dead ends.

Indeed, the phase separation of active layers of D–A bulk heterojunction solar cells is an intricate issue. A phase separation on the nanoscale is needed for efficient charge transport, yet macrophase separated polymer blends do not exhibit sufficient interfacial area for charge generation and exciton recombination will occur prior to charge separation [25–29]. In this respect, the molecular junction – as realized in molecular assemblies of D–A dyads, or in double cable polymers – enables intimate and stable mixing of donor and acceptor groups, whereas a polymer blend tends to minimize its interfacial area. Thus, these two extremes span the spectrum between highly intermixed donor and acceptor units on the one hand and a coarse phase separated morphology on the other hand. Consequently, something in between – a stable morphology with co-continuous domains of D and A few tens of nanometers thin – is sought after. Although this has been achieved quite successfully by accelerating and subsequent freezing in the demixing process of bulk heterojunction solar cells, the resulting morphologies are not stable and the domains are ill-defined. In contrast, the equilibrium microstructures of block copolymers are well-defined and can be tuned in size and shape when the molecular weight and the length of the individual blocks are varied [30]. Therefore, block copolymers that carry electronic functions hold great promise for the application in OPVs. Co-continuous morphologies suitable for photovoltaic active layers include cylindrical, lamellar or gyroidal phases. The construction of such microstructures from D–A block copolymers via self-assembly thus addresses the dilemma between the need for ordered transport pathways and sufficient optical absorption on a length scale that is commensurate with the exciton diffusion length [31]. Furthermore, techniques for preparing macroscopically aligned domains that are oriented perpendicular to the electrodes are well developed and have been successfully demonstrated using conventional block copolymers without electronic functions [32–34]. A graphical illustration of various morphological scenarios of active layers from blends and block copolymers is depicted in Scheme 1.

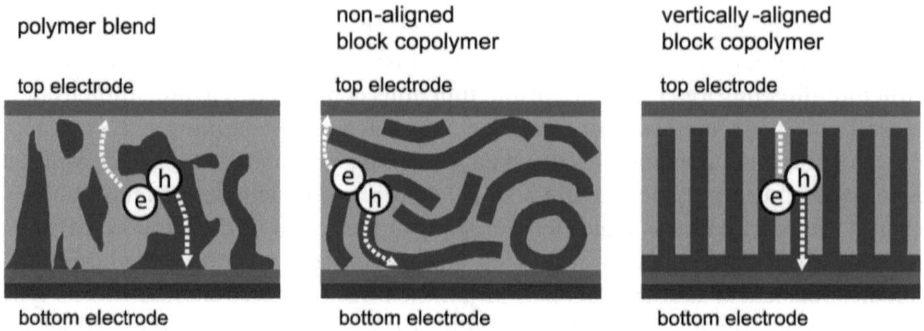

Scheme 1 Different possible donor–acceptor active layer morphologies in OPV devices. The *left drawing* depicts a macrophase separated polymer blend morphology with undefined domains. In the *middle* and at the *right*, schematic active layer morphologies of disordered and vertically aligned microphase separated block copolymer thin films are shown. *Dark and light grey domains* correspond to the donor and the acceptor phase, respectively

Whereas vertical alignment of cylindrical and lamellar block copolymers might be advantageous for improving the device performance, gyroidal films do not require alignment. Recently, an elegant example of this has been given by Snaith et al., who used a double sacrificial block copolymer to incorporate a gyroidal TiO_2 replicate into liquid electrolyte dye-sensitized solar cells [35]. However, the direct formation of the gyroid mesophase from a fully functionalized D–A block copolymer has not been realized up to now. The covalent connectivity of the donor and the acceptor blocks results in another positive advantage of these highly sophisticated polymer architectures: the equilibrium structure exhibits smaller domain sizes than polymer blends. Further crystallization of the two phases, concomitant with an increase in domain size and a decrease in interfacial area, can thus be excluded. This is an important issue concerning the morphological long-term stability of OPVs.

In contrast to conventional and commercially available block copolymers, block copolymers that carry electronically active blocks are relatively rare and the synthesis is challenging, since multi-step organic synthetic procedures have to be combined with one or more polymerization techniques. Further difficulties arise from the limited solubilities and the limited amount of material available from one batch, making the preparation of such materials tedious and time-consuming.

The first D–A block copolymers with suitable electronic properties for charge separation were synthesized by the group of Hadziioannou et al., using a conjugated poly(phenylene vinylene) (PPV) block as macroinitiator for the nitroxide mediated radical polymerization (NMRP) of a second styrenic coil block. This second segment was then converted into the acceptor block by attaching C_{60} molecules [36]. As pointed out by the authors, very strong interactions between fullerene units arising either from partial cross-linking [37] or from crystallization [38] possibly accounted for the lack of microdomains after C_{60} functionalization. In another study, Scherf et al. prepared D–A triblock copolymers from monobromo-terminated P3HT and dibromo-cyano substituted CN-PPVs via Yamamoto coupling [39]. Frechét et al. made use of ring opening metathesis polymerization (ROMP) to polymerize subsequently two macromonomers containing P3HT and fullerene. The products found application as compatibilizers in bulk heterojunction solar cells [40]. However, microphase separation was not demonstrated in any of these systems and only in two cases was a photovoltaic effect with solely the block copolymer as the active layer reported [38, 41].

The D and A moieties used in the majority of the above-mentioned synthetic approaches towards donor–acceptor block copolymers are conjugated polymers and fullerene derivatives. This makes the preparation very challenging and special attention has to be given to the introduction of appropriate solubilizing groups. A low weight fraction of, e.g. alkyl chains should render the polymer insoluble whereas a solubilizing group fraction that is too high will result in poor performance of the device since the amount of active material decreases. This is a general issue in such systems, becoming clearly visible in the case of polymers containing fullerene [42]. PBI as an alternative electron acceptor has been investigated to a lesser degree since the PCEs of solution-processed devices did not reach those of comparable polymer-fullerene cells [19, 43]. The main reason was seen in the uncontrolled crystallization of PBI, resulting in large one-dimensional crystals concomitant with poor morphological control [19]. Yet, suitable electronic properties and absorption in the visible range make this acceptor compound interesting for light harvesting applications and, apparently, interest in PBI for photovoltaic applications is reviving [20, 21, 44]. In addition, the chemical derivatization of the PBI core is feasible since the two distinct imide positions can be substituted independently without altering the electronic properties [45, 46]. Making use of these facts, Thelakkat et al. designed a highly soluble and polymerizable PBI derivative bearing a branched alkyl substituent at one imide position and a linear, acrylate-functionalized alkyl spacer at the other imide position. The incorporation of this monomer into block copolymers with poly(triphenylamine) as the donor yielded highly soluble D–A block copolymers exhibiting all important requirements for photovoltaic applications [47]. The valuable design and synthesis of this polymerizable electron conducting monomer marked the beginning of a variety of block copolymers with side-chain crystalline PBI blocks and donor blocks, being either amorphous or crystalline (Scheme 2).

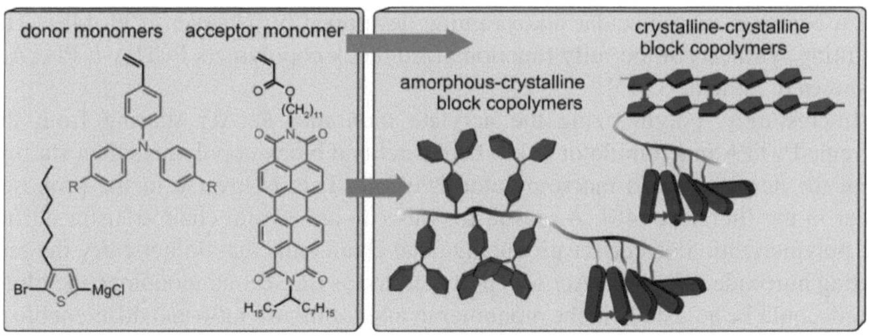

Scheme 2 Various donor and acceptor monomers can be combined to obtain block copolymers with amorphous or crystalline segments. In the *left box*, the polymerizable monomers are shown. On the *right*, the architectures of the resulting two main classes of D–A block copolymers are depicted: amorphous–crystalline and crystalline–crystalline block copolymers

The following sections summarize the work of our group concerning the synthesis, the morphology and the application of such D–A block copolymers in photovoltaic devices. The chapter is divided into two main sections. The first section covers amorphous–crystalline block copolymers, in which the donor function is taken care of by amorphous side-chain poly(triarylamine)s with different substituents. In the second section, very recent work on crystalline–crystalline block copolymers incorporating P3HT as the donor block is presented (Scheme 2). We try to picture comprehensively this interdisciplinary field by giving special emphasis to achievements as well as drawbacks in chemical synthesis. Furthermore, a broad and detailed characterization of the novel materials is given. Finally, the chapter finishes with results on the morphology of these complex materials, and the first promising results concerning the application in photovoltaic devices are presented.

2 Amorphous–Crystalline Donor Acceptor Block Copolymers

2.1 Synthesis and Characterization

With the advent of the polymerizable acceptor monomer perylene bisimide acrylate (PerAcr), an electronically eligible donor monomer and a suitable polymerization method had to be picked from the toolbox of organic and polymer chemistry. We chose vinyltriphenylamine (vTPA) as a starting point since the synthesis can be carried out in only one step and its energy levels form sufficient HOMO offset with PBI. This assured a sufficient driving force for efficient electron transfer from D to A [29]. In general, living or "quasi-living" polymerization methods were necessary to create the required block copolymer architectures. Additional criteria such as tolerance to functional groups and impurities as well as a metal-free nature were to be preferred. We therefore selected controlled radical polymerizations. Among the methods available, NMRP conformed to these preconditions [48]. Based on these building blocks, sequential NMRP of vTPA and PerAcr was performed using a common unimolecular alkoxyamine developed by Hawker et al. [49]. The resulting synthesis of the fully functionalized block copolymers PvTPA-*b*-PPerAcr is shown in Scheme 3 [47].

Interestingly, polymerizing the acrylate monomer PerAcr starting from the styrenic PvTPA macroinitiator led to better-defined block copolymers than starting from the acrylate-based macroinitiator PPerAcr. This is inverse to the proposed order in the literature [48]. A reason for this can arise from chain transfer during the polymerization of PerAcr producing dead chain ends that do not carry the mediating nitroxide. When PerAcr was polymerized as the second monomer, the block length could be adjusted via the monomer to macroinitiator ratio and different block copolymers with varying acceptor content were obtained.

In order to vary the HOMO offset at the D–A heterojunction as well as the hole mobility of the donor block, we also prepared substituted triphenylamine and tetraphenylbenzidine (TPD) monomers with two methoxy substituents [50, 51].

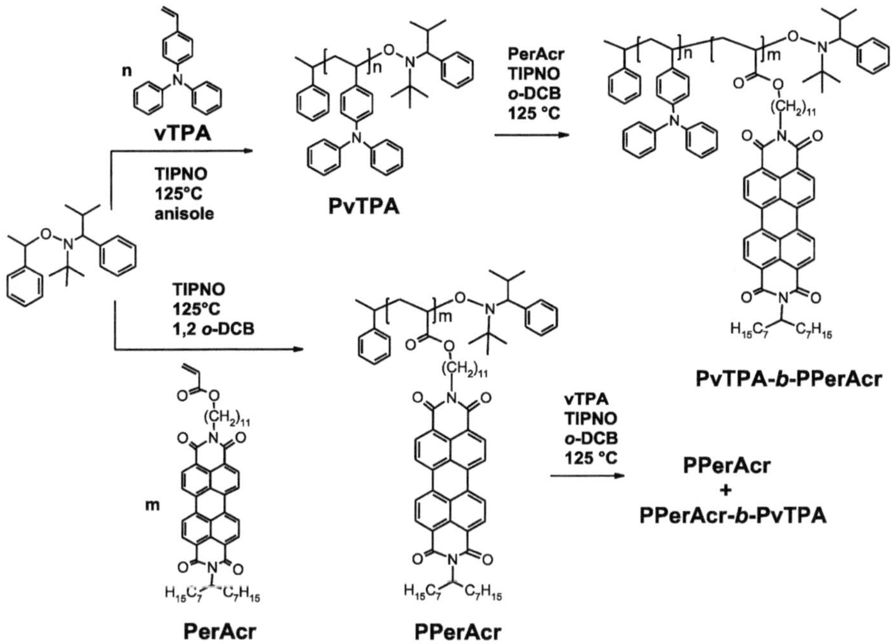

Scheme 3 Synthetic scheme of the two possible pathways towards PvTPA-*b*-PPerAcr block copolymers via NMRP. Using PvTPA as a macroinitiator for the NMRP of PerAcr gave monomodal block copolymers whereas starting from PPerAcr as macroinitiator, a mixture of PPerAcr and the corresponding PPerAcr-*b*-PvTPA block copolymer was obtained. *o-DCB ortho*-dichlorobenzene, *TIPNO* 2,2,5-trimethyl-4-phenyl-3-azahexane-3-oxyl

Scheme 4 shows the synthesis of bis(4-methoxyphenyl)-4′-vinylphenylamine (vDMTPA) and N,N'-bis(4-methoxyphenyl)-N-phenyl-N'-4-vinylphenyl-(1,1′ biphenyl)-4,4′-diamine (vDMTPD). The first monomer vDMTPA can be prepared from 4-bromostyrene and 4,4′-dimethoxydiphenylamine via a Buchwald–Hartwig amination in one step [52]. However, the formation of the side product bis(4-methoxyphenyl)-4′-vinylstyrylamine with a very similar retention factor made the purification of large monomer batches difficult (Scheme 4a). To circumvent this, a three-step synthetic procedure was chosen instead as outlined in Scheme 4b. Interestingly, vTPA gives a white powder after concentrating from solution whereas vDMTPA is viscous and difficult to handle, and therefore crystals of vDMTPA were grown from *n*-hexanes. The latter step was vital for an increase in the control of the polymerization and the common features of a living character were observed. The NMRP in anisole for a monomer-initiator-free nitroxide ratio of 200:1:0.1 resulted in low polydispersities between 1.1 and 1.2. The desired molecular weights of PvDMTPA were obtained by either varying the polymerization time or the initiator to monomer ratio. Figure 1a shows the linear increase of the experimental molecular weight with conversion, and Fig. 1b size exclusion chromatography (SEC) curves of different macroinitiators with molecular weights between 5 and 30 $\mathrm{kg\,mol^{-1}}$.

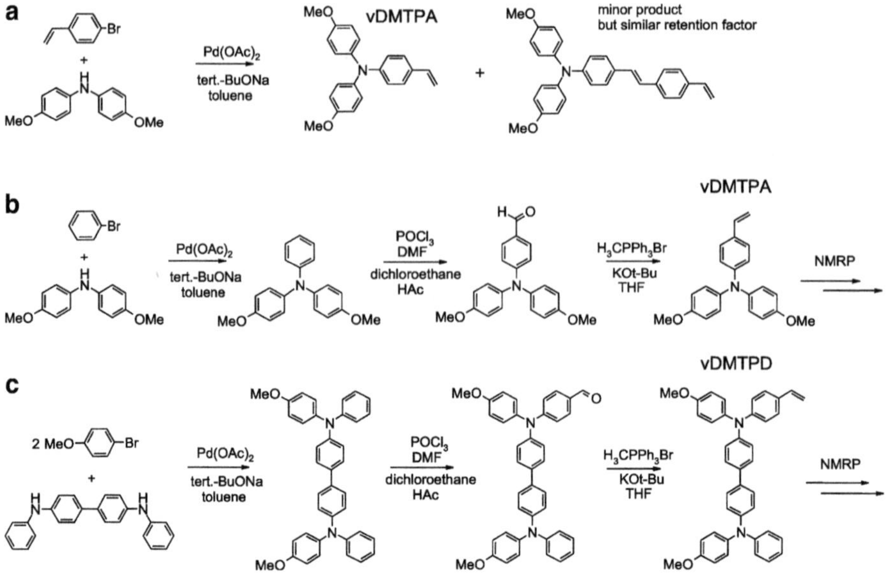

Scheme 4 **a** One-step synthesis of bis(4-methoxyphenyl)-4′-vinylphenylamine (vDMTPA) and formation of bis(4-methoxyphenyl)-4′-vinylstyrylamine. **b** Three-step procedure of vDMTPA. **c** Synthesis of *N*,*N*′-bis(4-methoxyphenyl)-*N*-phenyl-*N*′-4-vinylphenyl-(1,1′biphenyl)-4,4′-diamine (vDMTPD). Macroinitiator synthesis and subsequent preparation of block copolymers via NMRP were carried out analogously to Scheme 3

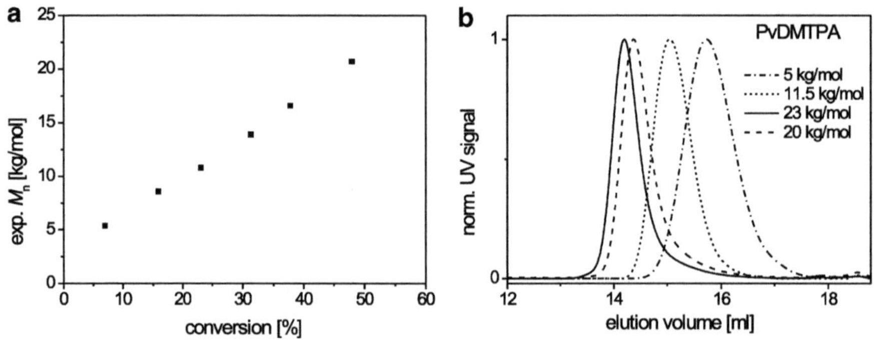

Fig. 1 **a** Linear increase of the experimental molecular weight of PvDMTPA vs conversion (determined by ^{1}H-NMR) for the NMRP of vDMTPA using a monomer-alkoxyamine-TIPNO ratio of 200:1:0.1. The extrapolated experimental molecular weight of 3.2 kg mol^{-1} at conversion = 0% arises from calibration with polystyrene standards. **b** SEC curves of the elution volume of different well-defined PvDMTPA macroinitiators

The block copolymer synthesis of PvDMTPA-*b*-PPerAcr was conducted analogously to the reactions with PvTPA as a macroinitiator (see Scheme 3). Note that block copolymer synthesis with PerAcr became more difficult with increasing molecular weight of the macroinitiator. Longer reaction times were necessary and the block copolymers generally exhibited broader molecular weight distributions. This might be due to the increased sterical hindrance of the active chain end with increasing molecular weight of the macroinitiator, retarding the polymerization of PerAcr and increasing the probability of side reactions such as chain end decomposition [53]. An illustration for this is given in Figs. 2a,b, where the SEC curves of two PvDMTPA macroinitiators with different molecular weights are shown together with their resulting block copolymers. Obviously, the SEC curve of the block copolymer in Fig. 2a (M_n PvDMTPA $= 9\,\text{kg mol}^{-1}$) is symmetric and no macroinitiator signal is found, whereas the curve of block copolymer in Fig. 2b is asymmetric due to remaining PvDMTPA (M_n PvDMTPA $= 18\,\text{kg mol}^{-1}$).

Finally, the concept of the amorphous side chain poly(triarylamines) was further extended to segments carrying TPDs. In general, the charge carrier mobility of TPDs is expected to be superior compared to simple TPAs [54, 55]. Two methoxy-groups were introduced to the TPD unit in the 4-position for the sake of solubility.

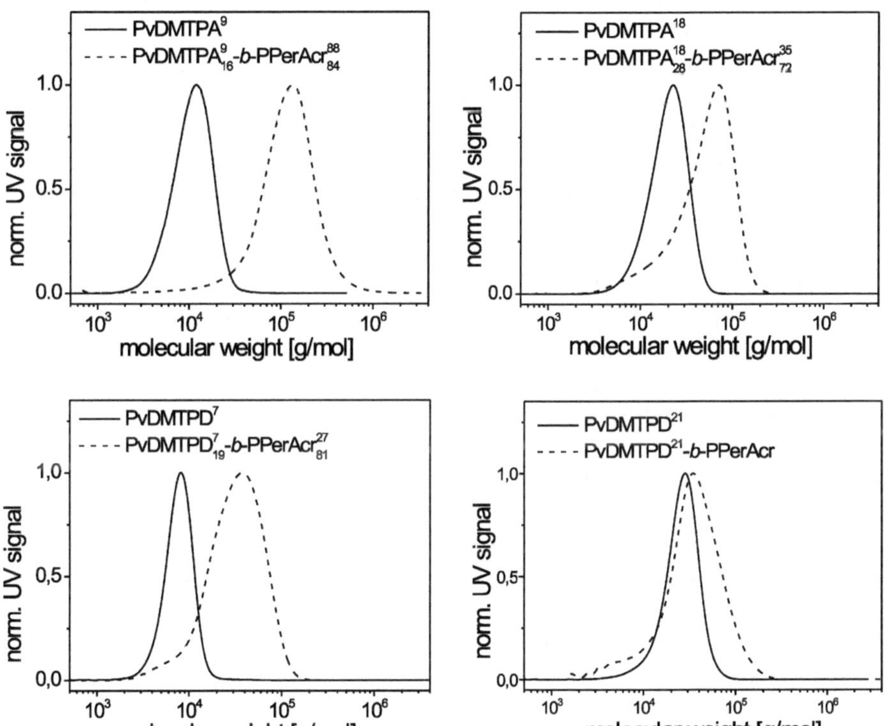

Fig. 2 SEC curves of PvDMTPA and PvDMTPD macroinitiators together with their resulting block copolymers PvDMTPA-*b*-PPerAcr and PvDMTPD-*b*-PPerAcr. PvDMTPA blocks with an M_n of 9 and $18\,\text{kg mol}^{-1}$ were used in (**a**) and (**b**), and PvDMTPD blocks with an M_n of 7 and $21\,\text{kg mol}^{-1}$ in (**c**) and (**d**), respectively. Subscripts denote the weight fractions. The first superscript denotes the molecular weight of the macroinitiator and the second one the molecular weight of the block copolymer in kilogram per mole

Such electron-rich groups also give rise to tuning of the HOMO level [56], providing an important tool for not only controlling the solubility but also the electronic properties. The monomer synthesis of *N*,*N*′-bis(4-methoxyphenyl)-*N*-phenyl-*N*′-4-vinylphenyl-(1,1′ biphenyl)-4,4′-diamine (vDMTPD) and the subsequent polymerization via NMRP is depicted in Scheme 3. Similar conditions already applied to the polymerizations of vDMTPA were used here and resulted in narrow-distributed and well defined macroinitiators PvDMTPD with polydispersities of ~1.2. However, polymerizations of vDMTPD had to be quenched at lower conversions, typically at ~30%, since the viscosity of the reaction mixture increased faster compared to PvDMTPA.

When using PvDMTPD as a macroinitiator for the polymerization of PerAcr, the success of the block copolymer synthesis generally was strongly influenced by the molecular weight of PvDMTPD. After the polymerization of PerAcr, a clear shift in the SEC curves towards smaller elution volumes was observed for macroinitiators with chain lengths up to 10 $kg\,mol^{-1}$, even though a shoulder remained at the elution volumes of the macroinitiator (Fig. 2c). This indicates the presence of non-functionalized PvDMTPD chains, arising either from termination reactions during vDMTPD polymerization or from chain termination at early stages during the initiation step of the synthesis of the second block. PvDMTPD macroinitiators with higher molecular weights between 15 and 20 $kg\,mol^{-1}$ could not successfully start the controlled polymerization of PerAcr. In such cases, the SEC traces of the products did not shift to smaller elution volumes, and peak broadening occurred instead (Fig. 2d). Apparently, the increasingly challenging synthesis of the second block with higher molecular weights of the macroinitiator already observed in the case of PvDMTPA pursued for PvDMTPD. The sterical hindrance and the electron richness of the PvDMTPD radical might be responsible for this trend. An illustration for this is given in Scheme 5, in which the general chain end composition mechanism of NMRP is depicted for PvDMTPD.

Indeed, the ^{1}H-NMR spectrum of the ethyl methyl ketone fraction of the product after PerAcr polymerization confirmed this assumption by featuring two doublets at 5.57 and 5.82 ppm, indicating the formation of the alkene-terminated PvDMTPD chain ends.

The issue of chain end degradation also becomes visible when PvDMTPD is used as a macroinitiator for the NMRP of commercially available monomers such as 4-vinylpyridine (4vP). However, due to the fast polymerization of 4vP, chain end degradation in the beginning of the polymerization is likely to be suppressed to a certain degree and high molecular weight PvDMTPD macroinitiators were successfully incorporated into block copolymers PvDMTPD-*b*-P4vP [57]. Mixtures of PvDMTPD-*b*-P4vP and inorganic CdSe:Te nanocrystals formed highly ordered lamellar block copolymer morphologies with the inorganic n-type particles located exclusively in the P4vP domains. Such inorganic nanocrystal:block copolymer hybrid materials are very interesting for photovoltaic applications, since they combine the different and various advantages of the single systems, namely high electron mobility, morphological control on a nanoscopic level and processability.

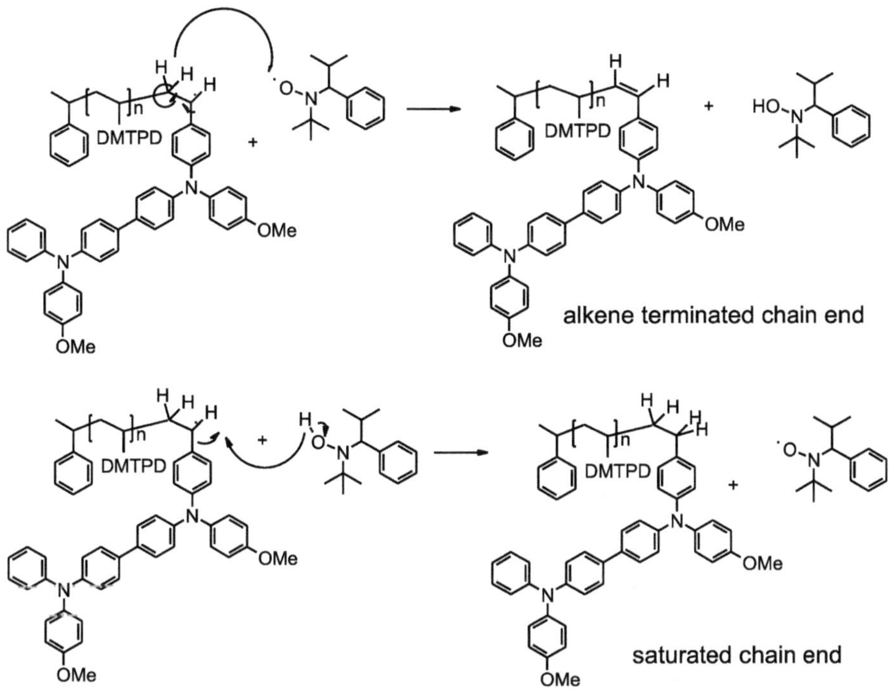

Scheme 5 Possible chain end degradation mechanisms during the NMRP of vDMTPD. Termination can occur during the polymerization of vDMTPD or when using PvDMTPD as a macroinitiator for the polymerization of the electron conducting monomer PerAcr

2.2 *Thermal Properties*

The thermal properties of the homopolymers and block copolymers were investigated by thermogravimetric analysis (TGA) and differential scanning calorimetry (DSC). These measurements determine thermal stabilities and phase transitions of the materials, information that is necessary for later device annealing procedures. In the case of block copolymers with crystalline segments, additional insight can be gained concerning crystal size, degree of crystallinity and mechanism of crystallization. Figure 3a shows the TGA curves of the two homopolymers PvDMTPA and PPerAcr and the block copolymer $PvDMTPA_{16}{}^{9}$-*b*-$PPerAcr_{84}{}^{88}$. Subscripts denote the weight fractions. The first superscript denotes the molecular weight of the macroinitiator and the second one the molecular weight of the block copolymer in kilograms per mole.

All three compounds are thermally stable and exhibit onset temperatures of 221 °C, 323 °C and 306 °C, respectively. Also, the thermal stability is sufficiently high, and decomposition occurs only above the temperature range of the phase transitions. Figure 3b shows the DSC traces of these compounds. A glass transition temperature T_g of 129 °C is observed for the amorphous PvDMTPA and, compared

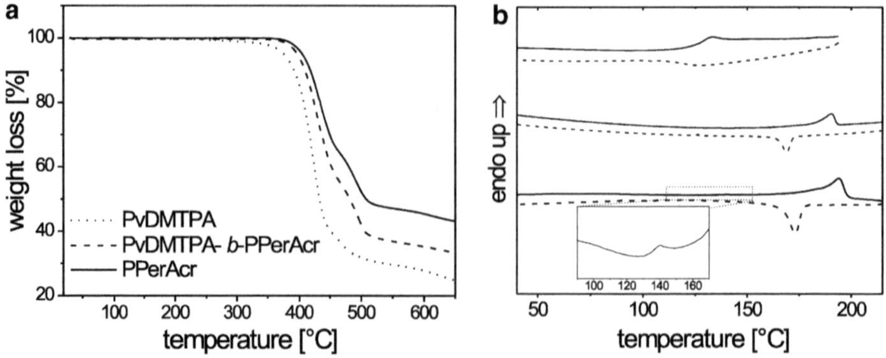

Fig. 3 **a** TGA curves of PvDMTPA, PPerAcr and PvDMTPA$_{16}{}^{9}$-*b*-PPerAcr$_{84}{}^{88}$. Curves were measured at $10\,\mathrm{K\,min^{-1}}$ under nitrogen. **b** DSC curves of the same polymers measured at $10\,\mathrm{K\,min^{-1}}$ in the case of PPerAcr and at $20\,\mathrm{K\,min^{-1}}$ in the case of PvDMTPA and PvDMTPA$_{16}{}^{9}$-*b*-PPerAcr$_{84}{}^{88}$. The second heating (*solid line*) and the second cooling (*dashed line*) cycle is shown; curves are offset for clarity

to PvTPA ($T_g = 150\,°\mathrm{C}$), is reduced by 21 °C due to the methoxy substituents in the *para* position. The PPerAcr block features a melting temperature T_m of 190 °C due to the side-chain crystallization. This side-chain crystallinity is caused by the strong π–π interactions of neighbouring PBI units. Both transitions are present in the block copolymer – even though the T_g is weak due to the small weight fraction of the amorphous block of only 16% (see inset of Fig. 3b) – and the temperatures of the transitions only deviate marginally compared to the homopolymers. The latter fact supports a microphase separation that leads to a crystalline PPerAcr phase in an amorphous donor matrix. The melting enthalpies of PPerAcr and PvDMTPA $_{16}{}^{9}$-*b*-PPerAcr$_{84}{}^{88}$ are 9.8 and $7.9\,\mathrm{J\,g^{-1}}$, respectively. In general, compared to the homopolymer, a decrease of the melting enthalpies of the PPerAcr block with decreasing block length was observed.

2.3 *Electronic and Optical Properties*

In this section the materials are investigated in terms of their energy levels, their UV–vis absorption profiles in film and their charge carrier mobilities. The HOMO and LUMO levels are estimated from cyclic voltammetry (CV) in solution and UV–vis spectroscopy. The charge carrier mobilities of PvTPA, PvDMTPA, PvDMTPD, and PPerAcr are extracted from measurements on organic field effect transistors (OFETs). CV measurements of PvTPA in solution feature an irreversible oxidation peak at 0.4 V vs ferrocene (Fc), giving a HOMO energy level of 5.2 eV (Fig. 4a,b) [51]. Irreversible oxidations are commonly observed for unsubstituted TPAs and the evolving new bands arise from dimerization products that are oxidized at lower potentials [58]. PvDMTPA is oxidized at 0.2 V vs Fc, corresponding to a HOMO level of 5.0 eV. The oxidation peak remains constant after several

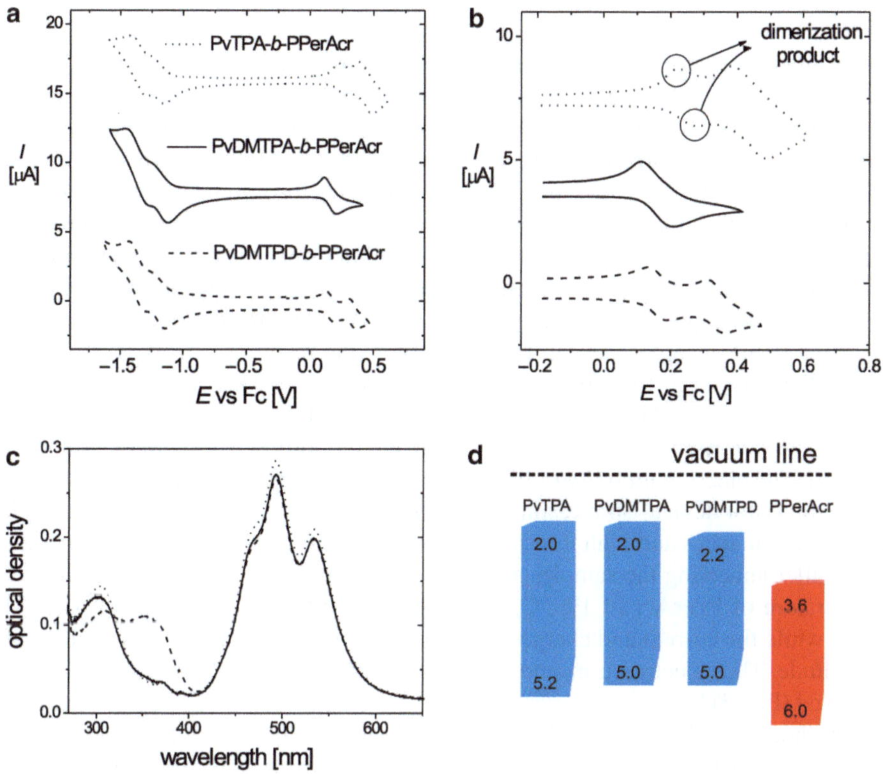

Fig. 4 **a** Cyclic voltammograms of PvTPA-b-PPerAcr (*dotted line*), PvDMTPA-*b*-PPerAcr (*solid line*) and PvDMTPD-*b*-PPerAcr (*dashed line*), showing the first and second reduction of PPerAcr and the oxidations of the poly(triarylamine)s. **b** Enlarged region of poly(triphenylamine) oxidation. All curves were measured in *o*-dichlorobenzene containing tetrabutyl ammonium hexafluorophosphate at 50 mV s^{-1} vs Fc (*Fc*: ferrocene). **c** Optical densities of films (70 nm) of PvTPA-*b*-PPerAcr (*dotted line*), PvDMTPA-*b*-PPerAcr (*solid line*) and PvDMTPD-*b*-PPerAcr (*dashed line*). Due to the similar content of PPerAcr in all block copolymers, the optical density between 400 and 600 nm is almost equal. **d** Schematic depiction of the estimated energy levels in electron volts. (Parts of this figure are reproduced with permission from [51]. Copyright Wiley-VCH Verlag GmbH & Co. KGaA)

cycles, demonstrating the reversibility of oxidation. In addition, the +M-effect of the methoxygroups contributes to the electrochemical stability of the radical cation [59]. Surprisingly, the first oxidation of PvDMTPD also occurs at 0.2 V vs Fc, resulting in the same HOMO level as PvDMTPA. The cyclic voltammogram of PPerAcr did not exhibit a clear oxidation peak, and therefore the monomer PerAcr was used for HOMO level determination. This compound showed an oxidation at 1.2 V vs Fc, corresponding to a HOMO level of 6.0 eV. The first reductions of PerAcr and PPerAcr both occurred at −1.2 eV. Thus, the position of the LUMO level is 3.6 eV, and is found to be independent of the molecular weight of PPerAcr. Generally, all events of oxidation and reduction were found to be independent of the molecular weight and the presence of the second block. The LUMO energy levels of the amorphous donor blocks were estimated from the CV results and the UV–vis absorbance in film. Figure 4c shows the absorption of the three block copolymers. The donor blocks exhibit the typical triarylamine absorption up to 400 nm and PPerAcr features three characteristic bands between 400 and 600 nm, corresponding to an aggregated state of the PBI moieties [60]. Since the absorption profile is sensitive to changes in the relative orientation of the PBI chromophores [61], we propose that stacking of PBI moieties occurs along one-dimensional stacks with a rotational offset, in a similar fashion to what has been shown for low molecular weight PBIs [17]. Figure 4d depicts the resulting energy levels of the three donor polymers with PPerAcr as the acceptor. As can be clearly seen, the resulting energy level offsets are sufficiently large for efficient charge separation in all the block copolymers, being 0.8–1.0 eV for ΔHOMO and ~1.5 eV for ΔLUMO [51].

For the determination of the charge carrier mobility, OFETs were produced [62, 63]. The electron mobility of PPerAcr depended strongly on the thermal history of the transistor device. Spin casting PPerAcr from chloroform yielded devices with weak performances and high threshold voltages around 20 V. This changed dramatically after annealing the samples at 210 °C for 60 min, which is above the melting temperature of PPerAcr of 190 °C. The threshold voltage dropped significantly to 6.8 V, while the current and charge carrier mobility both increased by two orders of magnitude. Thus electron transport mobilities of up to 1.2×10^{-3}cm^2 V^{-1} s^{-1} were achieved (Fig. 5) [62].

The hole mobilities of PvTPA, PvDMTPA and PvDMTPD were measured as 1×10^{-7}, 3×10^{-5} and 1×10^{-4} cm^2 V^{-1} s^{-1}, respectively, after spin casting from toluene. After thermal annealing the OFET devices above the glass transition temperature, mobility values of 3×10^{-5}, 5×10^{-5} and 1×10^{-4} cm^{-2} V^{-1} s^{-1}, respectively, were obtained [63]. Based on these results, we can derive several essential differences between PvTPA and PvDMTPA. First, the HOMO level shifts from 5.2 to 5.0 eV. Second, PvDMTPA is electrochemically stable because the two *para* positions are blocked and cannot give rise to dimerization reactions. Third,

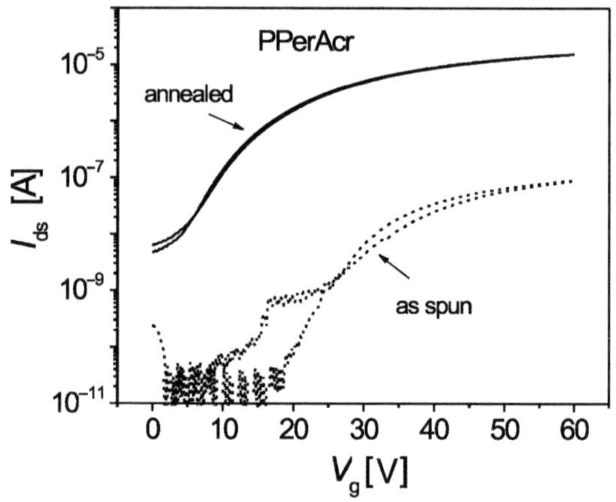

Fig. 5 Transfer characteristics of the PPerAcr homopolymer before and after annealing above the T_m (drain voltage 20 V). (Reproduced with permission from [62]. Copyright 2008 American Institute of Physics)

a slight improvement in the hole carrier mobility is observed, most likely due to the electron-rich methoxy substituents. The HOMO level position of PvDMTPA is maintained in PvDMTPD, which carries TPD moieties, and the charge carrier mobility of PvDMTPD is highest among the three donor polymers [63]. The trend of increasing hole carrier mobilities from PvTPA to PvDMTPA to PvDMTPD has also been confirmed using analysis of steady-state current–voltage characteristics from co-planar diode structures [64]. Thus, if incorporated into block copolymers with the acceptor polymer PPerAcr, this set of amorphous donor blocks is ideal for correlating the solar cell performance with important parameters such as energy levels, charge carrier mobility and morphology. By comparing these results with the morphological information from electron microscopy, certain structure-property relationships are established in the following section.

2.4 Morphology and Device Performance

In general, blends of amorphous donor polymers and PBI acceptor materials, being either small molecules or polymers, are not capable of undergoing a controlled phase separation. The crystallization of low molecular weight PBIs normally results in large and inhomogeneous crystals. Blends of PPerAcr and amorphous polymers will macrophase separate and form micron-sized domains. Block copolymers with one amorphous block and one side-chain crystalline PPerAcr block can solve this problem quite elegantly by confining PBI crystallization to nanometer-sized, either cylindrical or lamellar, domains [65]. This block copolymer approach is thus unique in controlling the size and shape of the PPerAcr acceptor microdomains, and can now be used to tune down the structural features to the range of the exciton diffusion length. Indeed, block copolymers with all the functionalities required for photovoltaic applications can outperform their analogous polymer blends, as was demonstrated recently for the case of PvTPA-*b*-PPerAcr (Fig. 6) [66].

The two TEM pictures of Fig. 6 unambiguously point out the advantage of the block copolymer in terms of creating a larger interfacial area between the donor and the acceptor phase (Fig. 6b). Macrophase separation, as it occurs in the blend film (Fig. 6a), is impeded in the block copolymer due to the covalent connectivity of the two blocks. The larger interface of D and A in the block copolymer film also gives rise to a higher degree of photoluminescence quenching (Fig. 6c) and finally leads to a sixfold external quantum efficiency (EQE) and to a tenfold improvement in the PCE. In the block copolymer film, more excitons can reach the D–A interface where they are separated into holes and electrons (Fig. 6d). Evidently, the structural features in Fig. 6b represent a non-equilibrium state after spin casting the block copolymer from chloroform solutions. If at all, the domains seem to be oriented rather parallel than perpendicular to the substrate. It has to be elucidated how such a block copolymer thin film morphology can be aligned vertically. Yet the large

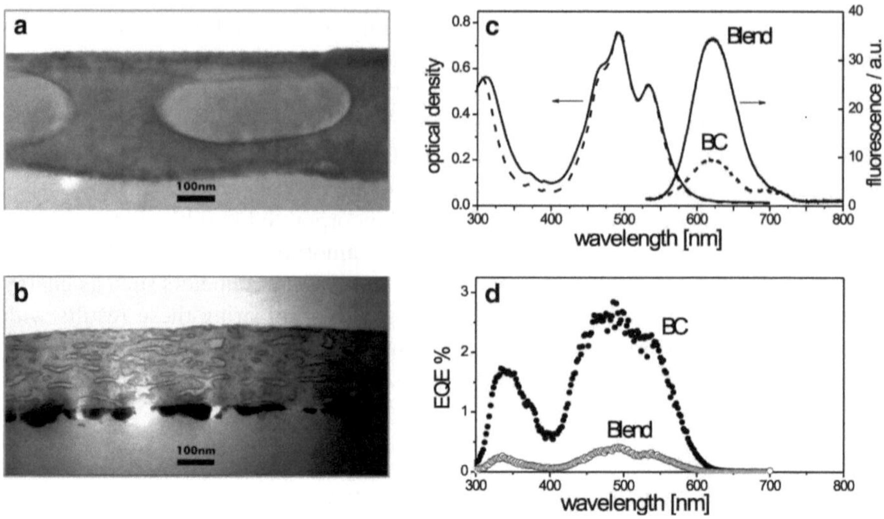

Fig. 6 **a** Cross sectional TEM of active layers of a PvTPA:PPerAcr blend and **b** of a PvTPA-*b*-PPerAcr block copolymer (BC). **c** Optical density and photoluminescence quenching. **d** External quantum efficiency (EQE) of these same films. The films are stained with RuO_4, bright regions in (**a**) and (**b**) are due to PPerAcr domains. (Reproduced with permission from [66]. Copyright Wiley-VCH Verlag GmbH & Co. KGaA)

improvement of the block copolymer device demonstrates the huge potential this approach offers and further enhancement is expected by optimizing processing and post-production treatments. However, improvements in device performance are also effective at the materials synthesis stage. A recent study of Haque et al. has emphasized the importance of the segment lengths to morphology and device performance in PvTPA-*b*-PPerAcr, even when the same type of nanostructure is obtained with different degrees of polymerization [29]. Furthermore, derivatizing the triphenylamine core of PvTPA with electron-rich substituents has led to considerable device improvements. In the following, the complex interplay between the HOMO level offset, the hole carrier mobility, the morphology and the device performance will be comprehensively outlined.

The morphologies of the block copolymers containing the different hole conductors PvTPA, PvDMTPA and PvDMTPD were first investigated by transmission electron microscopy of bulk samples [51]. For this purpose, the polymers were thermally annealed above their melting temperatures, cooled down at $10\,K\,min^{-1}$, embedded into epoxy resin and microtomed ($\sim$50 nm). Subsequent staining was carried out using RuO_4, which preferentially reacts with PPerAcr. The staining time turned out to be an important parameter. Short treatment with RuO_4 vapour was necessary to generate a sharp contrast between the two phases whereas staining times exceeding 30 min resulted in broader structures. The observed effect of initial boundary staining might be due to preferential staining of RuO_4 with the amorphous PPerAcr that surrounds the crystalline PPerAcr

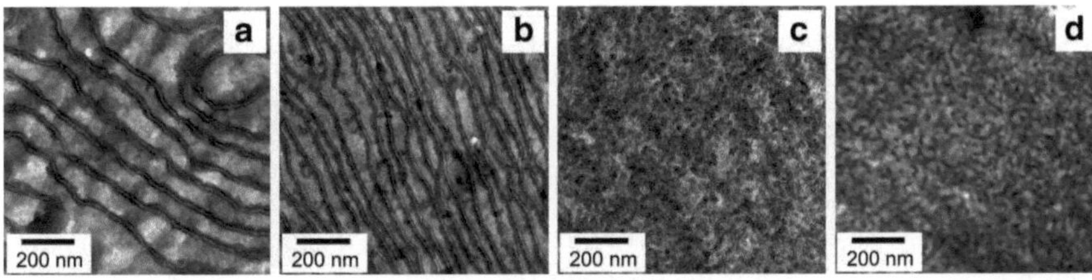

Fig. 7 TEM pictures of thermally annealed bulk samples of **a** $PvTPA_{14}{}^{16}$-*b*-$PPerAcr_{86}{}^{26}$, **b** $PvDMTPA_{28}{}^{18}$-*b*-$PPerAcr_{72}{}^{35}$, **c** $PvDMTPA_{16}{}^{9}$-*b*-$PPerAcr_{84}{}^{88}$ and **d** $PvDMTPD_{19}{}^{7}$-*b*-$PPerAcr_{81}{}^{26}$. *Dark domains* represent PPerAcr and the *brighter ones* the amorphous donor blocks. The samples were annealed above their melting temperatures, embedded into epoxy resin, microtomed and stained with RuO_4. (Reproduced with permission from [51]. Copyright Wiley-VCH Verlag GmbH & Co. KGaA)

phase. Such a phenomenon has been reported for semi-crystalline block copolymers poly(ethylene)-*b*-poly(vinylcyclohexane) [67]. Figure 7 shows the TEMs of PvTPA-*b*-PPerAcr, PvDMTPA-*b*-PPerAcr and PvDMTPD-*b*-PPerAcr.

Figure 7a shows block copolymer $PvTPA_{14}{}^{16}$-*b*-$PPerAcr_{86}{}^{26}$ with a donor block of $\sim$16 kg mol^{-1}, a PPerAcr weight fraction of 86% and an overall molecular weight of $\sim$26 kg mol^{-1}. The microphase separated material exhibits crystalline PPerAcr lamellae (dark) in an amorphous PvTPA matrix (brighter). A similar morphology is observed from $PvDMTPA_{28}{}^{18}$-*b*-$PPerAcr_{72}{}^{35}$ in Fig. 7b. Note that the term "wire-like" used earlier [50, 51] may not reflect the real morphology, and will be replaced by "lamellar". The structures in Fig. 7c,d are substantially different compared to Fig. 7a,b. Worms or fibrils of PPerAcr appear in the micrographs of both polymers, $PvDMTPA_{16}{}^{9}$-*b*-$PPerAcr_{84}{}^{88}$ (Fig. 7c) and $PvDMTPD_{19}{}^{7}$-*b*-$PPerAcr_{81}{}^{26}$ (Fig. 7d). The morphologies in Fig. 7a,b are of particular interest for photovoltaic applications since both domains, donor as well as acceptor, percolate through the whole bulk sample. Since the molecular weight of the donor block is 16–18 kg mol^{-1} in Fig. 7a,b and 7–9 kg mol^{-1} in Fig. 7c,d, the length of the amorphous segment seems to decide on the formation of a lamellar morphology if the PPerAcr weight fraction lies between 72% and 86%. In order to correlate these structures with the PV device performance, solar cells were constructed from the same polymers [51]. Note that the HOMO level offset is 1.0 eV in Fig. 7a and 0.8 eV in all the other polymers and that the hole carrier mobility increases from PvTPA (Fig. 7a) to PvDMTPA (Fig. 7b,c) to PvDMTPD (Fig. 7d). Indium tin-oxide (ITO) substrates covered with PEDOT:PSS were used for the preparation of photovoltaic devices. The active layer was spin cast from chloroform solutions (0.5 wt%) and the devices were completed with aluminium as top electrode. The current density–voltage (*J*–*V*) characteristics of these four polymers are shown in Fig. 8a.

We first compare the *J*–*V*-characteristics of the two block copolymers with PvTPA and PvDMTPA as donor, whose bulk morphologies are quite similar, as shown in Fig. 7a,b. The J_{SC} in the device made of $PvDMTPA_{28}{}^{18}$-*b*-$PPerAcr_{72}{}^{35}$ (*squares*) amounts to 1.14 mA cm^{-2} which is a fivefold increase compared to 0.23 mA cm^{-2} in $PvTPA_{14}{}^{16}$-*b*-$PPerAcr_{86}{}^{26}$ (*diamonds*). This indicates that D–A

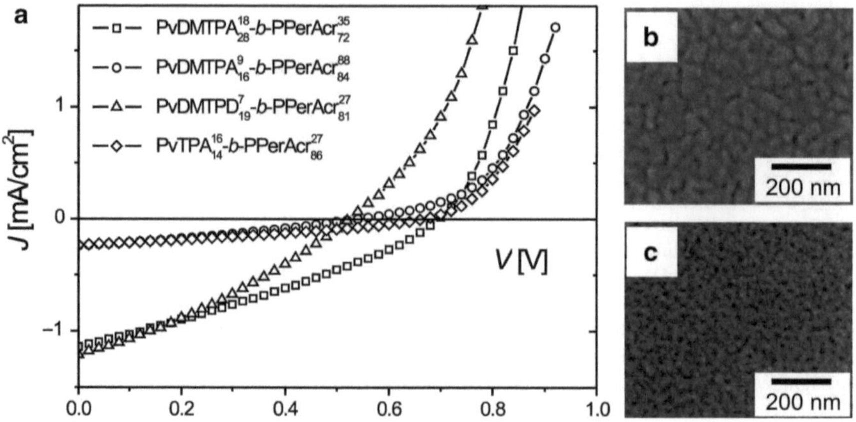

Fig. 8 **a** *J–V* -characteristics of block copolymers with different donors and different molecular weights. **b** SEM picture of the surface of the device made of $\text{PvDMTPA}_{28}{}^{18}$-*b*-$\text{PPerAcr}_{72}{}^{35}$ solar cells. **c** SEM picture of the surface of the device made of $\text{PvDMTPA}_{16}{}^{9}$-*b*-$\text{PPerAcr}_{84}{}^{88}$. *Bright domains* represent PPerAcr and the *darker ones* the amorphous donor blocks

charge transfer occurs more efficiently in the block copolymer with PvDMTPA than PvTPA as hole conductor, which is caused by the higher D/A HOMO offset (Fig. 4d). From the higher slope of the *J–V*-curve of $\text{PvDMTPA}_{28}{}^{18}$-*b*-$\text{PPerAcr}_{72}{}^{35}$ (*squares*) at open circuit conditions, it can be concluded that charges are injected more easily from PvDMTPA (HOMO = 5.0 eV) than from PvTPA (HOMO = 5.2 eV) into ITO. The maximum photovoltage that can be achieved at the D–A interface is given by HOMO_D–LUMO_A, implying a loss of photovoltage if HOMO_D is shifted to higher values. But this negative effect is compensated in $\text{PvDMTPA}_{28}{}^{18}$-*b*-$\text{PPerAcr}_{72}{}^{35}$ (*squares*) due to an improvement in the hole transfer efficiency and, as a result, the V_OC of both devices are nearly the same (670 and 690 mV).

In the following, the impact of morphology on the *J–V*-characteristics of PvDMTPA-*b*-PPerAcr block copolymers is elucidated. Block copolymer $\text{PvDMTPA}_{28}{}^{18}$-*b*-$\text{PPerAcr}_{72}{}^{35}$ (*squares*) self-assembles into a lamellar morphology and $\text{PvDMTPA}_{16}{}^{9}$-*b*-$\text{PPerAcr}_{84}{}^{88}$ (*circles*) exhibits shorter PPerAcr structures (see Fig. 7b,c). The PCE of the device with the lamellar block copolymer is improved fivefold compared to $\text{PvDMTPA}_{16}{}^{9}$-*b*-$\text{PPerAcr}_{84}{}^{88}$ (*circles*). Both parameters, the J_SC and the V_OC, rise from 0.24 to 1.14 mA cm^{-2} and from 530 to 690 mV, respectively. Since the HOMO level is 5.0 eV in both block copolymers and the charge carrier mobility is not altered, the difference in morphology must be responsible for the change in device performance. Also, the same amount of light is absorbed by the two block copolymers, as indicated by the same optical density of the active layers between 400 and 600 nm. $\text{PvDMTPA}_{28}{}^{18}$-*b*-$\text{PPerAcr}_{72}{}^{35}$ (*squares*) and $\text{PvDMTPA}_{16}{}^{9}$-*b*-$\text{PPerAcr}_{84}{}^{88}$ (*circles*) vary mainly in their hole conductor block length, resulting in different volume nanostructures. $\text{PvDMTPA}_{28}{}^{18}$-*b*-$\text{PPerAcr}_{72}{}^{35}$ (*squares*) shows a lamellar morphology and $\text{PvDMTPA}_{16}{}^{9}$-*b*-$\text{PPerAcr}_{84}{}^{88}$ (*circles*) exhibits shorter structures. As a consequence, charge percolation should be more favourable in the former case. This indicates that not only an adaptation of the energetic levels but also the generation of longer and defined charge percolation pathways are required for improving the efficiency of a bulk heterojunction solar cell. Albeit the correlation of thermally annealed bulk structures with the device performance of active layers spun from chloroform can rationalize these findings, such a comparison is critical in general. In order to provide information regarding the real device morphologies, scanning electron microscopy (SEM) was done on both PvDMTPA-*b*-PPerAcr devices after spin casting (Fig. 8b,c). In SEM, the contrast is inverted compared to TEM images. Now, the bright domains represent PPerAcr and the dark regions can be assigned to PvDMTPA. The surface of $PvDMTPA_{28}{}^{18}$-*b*-$PPerAcr_{72}{}^{35}$ (*squares*) exhibits large and flaky domains of PPerAcr whereas the structural features of $PvDMTPA_{16}{}^{9}$-*b*-$PPerAcr_{84}{}^{88}$ (*circles*) are smaller in terms of both, the donor as well as the acceptor domains size. On the one hand, this corroborates the picture of a better charge percolation in the device made of $PvDMTPA_{28}{}^{18}$-*b*-$PPerAcr_{72}{}^{35}$ (*squares*) and, on the other hand, indicates that a domain size that is too small leads to inefficient charge transport and increased recombination rates.

Finally, we consider the effect of a PvDMTPD segment carrying efficient TPD hole transport units and compare the device performance with the block copolymers made from PvDMTPA macroinitiators. The HOMO energy levels in PvDMTPD and PvDMTPA both are 5.0 eV so that any differences in device performance can be ascribed to changes in morphology or charge carrier mobility. Choosing $PvDMTPA_{16}{}^{9}$-*b*-$PPerAcr_{84}{}^{88}$ (*circles*) and $PvDMTPD_{19}{}^{7}$-*b*-$PPerAcr_{81}{}^{26}$ (*triangles*) for this comparison allows to exclude large effects of morphology, since their block lengths and compositions are similar (see Fig. 7c,d). In the device made of the PvDMTPD block copolymer, the improvement in the $J_{SC}(1.21\,\mathrm{mA\,cm^{-2}})$ is fivefold, whereas the open circuit voltage is 0.53 V in both cases. This result is encouraging since the short circuit current is highest among the four block copolymer devices, although the TEM nanomorphology suggests the presence of non-optimal worm-like structures. Therefore the improved current can be attributed to the higher charge carrier mobility of TPDs [63, 64]. Interestingly, devices made of block copolymers that exhibit lamellar nanostructures in the bulk (Fig. 7a,b) give higher open circuit voltages (0.67 and 0.69 V, respectively) than those with active layers from $PvDMTPA_{16}{}^{9}$-*b*-$PPerAcr_{84}{}^{88}$ (*circles*) and $PvDMTPD_{19}{}^{7}$-*b*-$PPerAcr_{81}{}^{26}$ (*triangles*) (Fig. 7c,d). We therefore believe that a block copolymer such as PvDMTPD-*block*-PPerAcr can be designed to obtain lamellar domains, in which the synergetic effects of an improved hole transfer efficiency, an acceptable charge carrier mobility and a desired morphology can lead to higher PCEs.

However, another approach for improving the charge carrier mobility is to employ conjugated, semi-crystalline polymers. Here, a further advantage is the extended absorption in the visible range. These issues are addressed in the next chapter which is concerned with crystalline–crystalline block copolymers comprised of poly(3-hexythiophene) and PPerAcr.

3 Crystalline–Crystalline Donor Acceptor Block Copolymers

3.1 Synthesis and Characterization

The objective of synthesizing block copolymers with P3HT and PPerAcr involves the combination of two polymerization methods, Grignard Metathese Polymerization (GRIM) and NMRP, in a straightforward fashion. Poly(3-alkylthiophenes) prepared via the GRIM method were first incorporated into block copolymers by McCullough et al. using polymeranalogous reactions to generate P3HT macroinitiators for the atom transfer radical polymerization (ATRP) of a second coil block [68, 69]. Later on, this concept was extended to P3HT macroinitiators for NMRP and reversible addition fragmentation termination polymerization (RAFT) [70]. The preparation of these macroinitiators included four steps and, therefore, we aimed at a one-pot procedure. Protocols for the in situ introduction of different endgroups were readily available from McCullough et al. [71], giving an obvious pathway for a straightforward synthetic methodology towards P3HT block copolymers in two steps (Scheme 6) [72].

The in situ introduction of the alkoxyamine at the end of the P3HT chain (capping efficiencies between 40% and 85%) was verified by ^{1}H-NMR. The success of this reaction sensitively depended on the fairly complex polymerization conditions. McCullough et al. found mixtures of mono- and dicapped P3HT species, depending on the type of Grignard used [71], whereas we could only detect small amounts of dicapped P3HT macroinitiators. A mixture of mono- and dicapped macroinitiators leads to a mixture of diblock and triblock copolymers, which cannot be separated afterwards by simple extraction methods. This complicates characterization and can be disadvantageous regarding the self-assembly of the material. To gain additional insight into the various termination reactions of the GRIM during the in situ endcapping with the alkoxyamine, the endgroups were analyzed using matrix-assisted laser

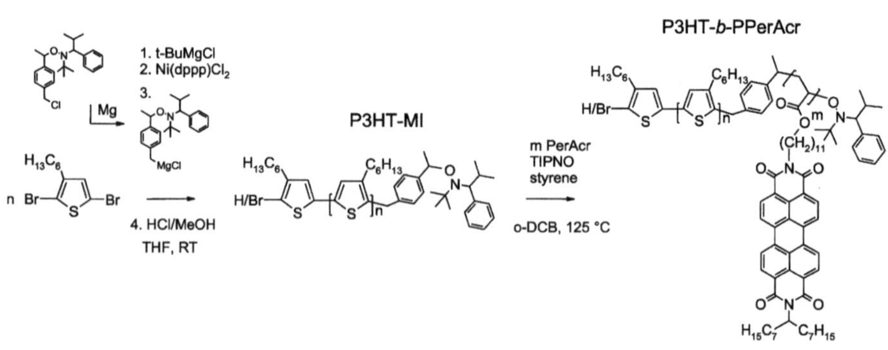

Scheme 6 One-pot synthesis of P3HT-MI macroinitiators for NMRP using the McCullough method followed by in situ endcapping with the Grignard derivative of a common alkoxyamine initiator. Starting from these macroinitiators, the acceptor monomer PerAcr is polymerized to give fully functionalized, double-crystalline block copolymers P3HT-*b*-PPerAcr

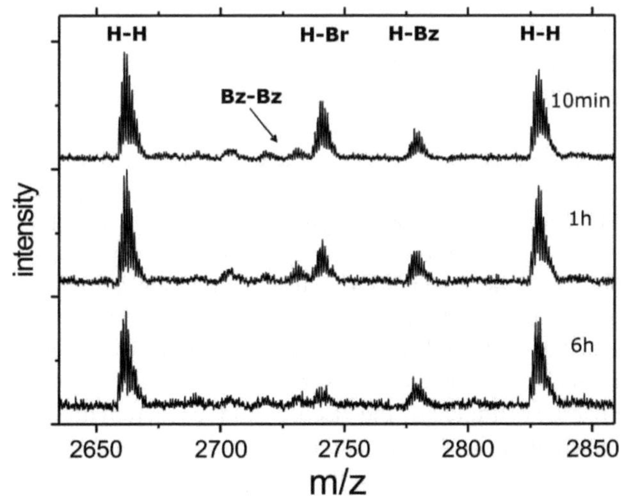

Fig. 9 Evolution of P3HT endgroups during the encapping reaction with the Grignard-functionalized alkoxyamine as revealed by MALDI-TOF. The 16-mer is shown. *Bz*: benzylic fragment of the alkoxyaminc

desorption ionization time-of-flight (MALDI-TOF) as a function of endcapping time. For this purpose, aliquots were withdrawn after adding an eightfold excess of the alkoxyamine endcapper, quenched with hydrochloric acid, and analyzed by MALDI-TOF [73]. The results are summarized in Fig. 9 and show the engroups of a P3HT-alkoxyamine macroinitiator after 10 min, 1 h and 6 h of encapping time.

The alkoxyamine does not survive the MALDI ionization procedure and only P3HT chains with the benzylic fragment (Bz) are observed (similar to mass spectrometry of the low molecular weight alkoxyamine). A quantitative evaluation of the MALDI-TOF peak heights is critical since the peak intensities do not represent proportionally the amount of the respective species. Qualitatively, P3HTs with hydrogen–hydrogen (H–H), hydrogen–bromine (H–Br), hydrogen–alkoxyamine (H–Bz) and alkoxyamine–alkoxyamine (Bz–Bz) endgroups are found. On comparing the different spectra, most of the alkoxyamine endcapping occurs during the first few minutes after adding the endcapper and increases only slightly afterwards. The overall degree of alkoxyamine endcapping is small in this case (~40% by ^{1}H-NMR), and despite the presence of further H–Br chain ends, these are not coupled to the initiator. Rather, the signal of the H–Br ends decreases, concomitant with an increase of the signal of the H–H endgroups. This indicates that, despite the presence of Grignard-functionalized alkoxyamine, H–Br-terminated chains undergo a side-reaction which converts them into H–H-terminated chains. A general strategy for obtaining higher yields of alkoxyamine-terminated P3HT is therefore to use a larger excess of endcapper. When an excess of alkoxyamine around 15 eq (with respect to the catalyst) was added, the degree of endcapping increased from 40% to up to 80%. However, a complete endcapping of all P3HT chains with alkoxyamine seems unlikely as long as degradation of bromine-terminated chains occurs during the GRIM. We believe that the simple and straightforward one-pot procedure compensates for incomplete encapping and, depending on the second segment, the non-functionalized P3HT can be removed afterwards via Soxhlet extraction. More critical is the presence of dicapped macroinitiators leading to triblock copolymers, since the separation of triblock and diblock copolymers is not possible by extraction methods. Using the in situ endcapping method presented here, the fraction of P3HT with two alkoxyamine groups is small and, as will be shown later, the polydispersities of the resulting block copolymers are sufficiently low (1.2–1.5).

In the second step, the P3HT-MI was used to polymerize PerAcr (Scheme 6). The segment length of the PPerAcr can be well-adjusted by the reaction time as well as the ratio [P3HT-MI] to [PerAcr]. Similar reaction conditions applied earlier for block copolymer synthesis with PerAcr were successfully used here. The yield of the polymerization is limited by the viscosity of the reaction mixture; the reaction is typically stopped after 30–40% conversion. Purification of the products can simply be achieved by Soxhlet extraction, thereby removing non-functionalized P3HT and monomer PerAcr. P3HT-MI macroinitiators with different molecular weights were synthesized and incorporated into block copolymers with PPerAcr, as the P3HT molecular weight determines the performance of solar cells to a large extent [74]. The size exclusion chromatograms (SEC) of the block copolymers P3HT-*b*-PPerAcr, hereafter referred to as BC 16, 17, 21, 25 and 30, and their corresponding macroinitiators P3HT-MI 9 and P3HT-MI 17, are shown in Fig. 10 and demonstrate the successful block copolymer synthesis.

The molecular weights of BC 16, 17, 21, 25 and 30 are 16.1, 16.9, 20.6, 24.8 and 29.5 $\mathrm{kg\,mol^{-1}}$, respectively, and the polydispersity indices are fairly low, between 1.15 and 1.31. Only for PPerAcr weight fractions as high as 80% in BC 25

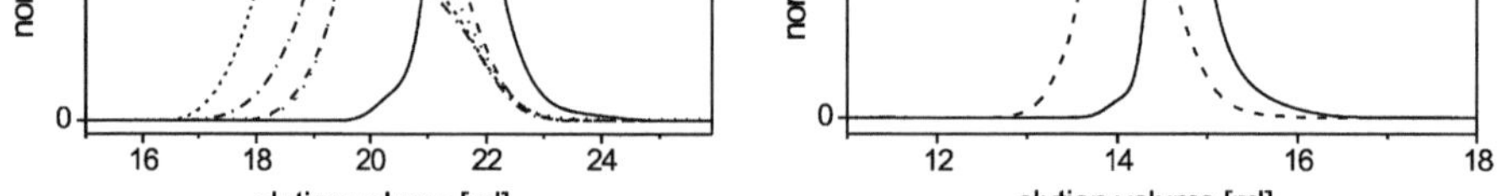

Fig. 10 **a** Size exclusion chromatography (SEC) of the macroinitiator P3HT-MI 9 (M_n = 8.9 $\mathrm{kg\,mol^{-1}}$) and the four block copolymers P3HT-*b*-PPerAcr BC 16, 17, 21 and 25 (M_n = 16.1, 16.9, 20.6 and 24.8 $\mathrm{kg\,mol^{-1}}$, respectively). (Reproduced with permission from [72]. Copyright Wiley-VCH Verlag GmbH & Co. KGaA.) **b** SEC curves of macroinitiator P3HT-MI 17 with a high molecular weight of 17.0 $\mathrm{kg\,mol^{-1}}$ and a resulting block copolymer P3HT-*b*-PPerAcr BC 30 (M_n = 29.5 $\mathrm{kg\,mol^{-1}}$). (Reproduced with permission from [75]. Copyright (2009), American Institute of Physics.) The curves were measured in THF containing 0.25 wt% tetrabutylammonium bromide in (**a**) and pure THF was used as the eluent in (**b**)

Table 1 Molecular weights, polydispersity indices, compositions and thermal properties of homo- and block copolymers. The molecular weights were determined via SEC using polystyrene calibrations, the compositions via ^{1}H-NMR and the thermal properties from DSC measurements. The melting enthalpies are normalized to their weight fractions

Polymer	$M_{n,\mathrm{P3HT}}$ (kg mol^{-1})	$M_{n,\mathrm{overall}}$ (kg mol^{-1})	PDI	wt% PPerAcr	T_{m1} (°C)	T_{m2} (°C)	T_{c1} (°C)	T_{c2} (°C)	$\Delta H_{\mathrm{m,P3HT}}$ (J g^{-1})
PPerAcr	–	23	1.71	100	191	–	169	–	–
P3HT-MI 9	8.9	–	1.12	0	208	–	180	–	13.1
P3HT-MI 17	17.0	–	1.12	0	223	233[a]	192	–	16.9
BC 16	8.9	16.1	1.25	55.7	190	211	148	162	10.3
BC 17	8.9	16.9	1.24	59.7	191	211	148	163	8.8
BC 21	8.9	20.6	1.31	73.7	202[b]	–	172[b]	–	–[b]
BC 25	8.9	24.8	1.53	81.4	206[b]	–	179[b]	–	–[b]
BC 30	17.0	29.5	1.15	54.9	204	244	178[c]	–[c]	15.4

[a] The two melting points of P3HT-MI 17 are probably due to a smectic liquid crystalline behaviour (see [76])
[b] A single melting and crystallization peak appears
[c] The recrystallization of BC 30 shows a shoulder at ~176°C beside the main peak at 178°C

does the PDI increase to 1.53. For the preparation of active layers in bulk heterojunction solar cells, the donor acceptor blend composition commonly lies between 1:1 and 1:4, depending on the polymer used. The PPerAcr weight fractions in our set of block copolymers exactly matches with these values, spanning the range between ~50 wt% (BC 16) and ~80 wt% (BC 25). Similarly to our observations regarding the increased difficulties of PerAcr polymerization with increasing M_{n} of poly(triarylamine) (see Fig. 2), PPerAcr blocks were more difficult to synthesize starting from the higher molecular weight macroinitiator P3HT-MI 17. This was manifested by prolonged reaction times and higher monomer to macroinitiator ratios that were necessary to obtain the same compositions compared to P3HT-MI 9. All parameters regarding the molecular weight, PDI and composition are shown in Table 1.

3.2 Thermal Properties

The double-crystalline character of P3HT-*b*-PPerAcr was investigated by DSC (Fig. 11) [72, 75]. The macro-P3HT-9 exhibited a melting point (T_{m}) at 211 °C, whereas the homopolymer PPerAcr melted at 191 °C. In BC 16, the two T_{m}s observed at 190 °C and 211 °C were ascribed to melting of PPerAcr and P3HT domains, respectively. The cooling curve of BC 16 showed two exotherms at 162 °C and 148 °C, that are ascribed to the recrystallizations (T_{c}) of PPerAcr and P3HT, respectively. The macro-P3HT 17 showed two T_{m}s at 223 °C and 233°C. The observation of a second melting peak probably is due to a smectic liquid crystalline behaviour [76]. BC 30 exhibited two T_{m}s at 204 °C and 244 °C in the heating curve,

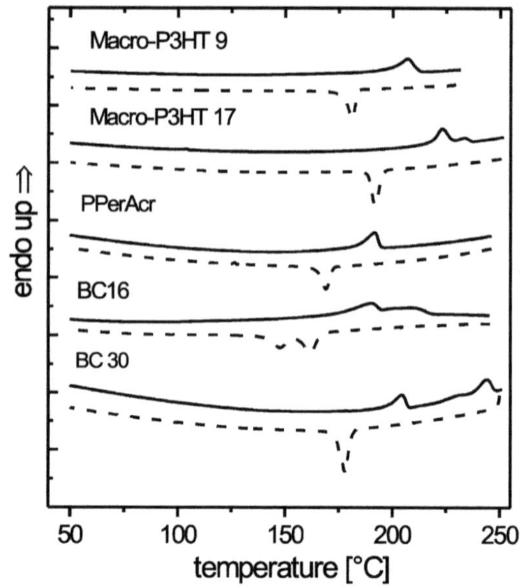

Fig. 11 Differential scanning calorimetry of P3HT-*b*-PPerAcr, the corresponding PPerAcr homopolymer, and the P3HT-macroinitiators. The second heating (*solid line*) and the second cooling curve (*dashed line*) is shown; curves were measured at $10\,\mathrm{K\,min^{-1}}$ under nitrogen, and are offset for clarity

but only one T_c at 178 °C in the cooling curve. This behaviour is caused by the higher segment length of P3HT, which shifted the melting and recrystallization temperatures of P3HT towards higher values. As a result, the distance between the two T_ms was larger in BC 30 compared to BC 16, whereas the distance between the two T_cs in BC 16 became smaller with increasing molecular weight of P3HT, and the two peaks appeared as one transition in the cooling curve of BC 30. The higher melting point of the P3HT block in BC 30 points out larger P3HT crystals. We also note that the degree of P3HT crystallinity was higher in BC 30 compared to BC 16 as indicated by the melting enthalpies ΔH_m of the P3HT melting peak normalized to the weight fraction. ΔH_m of BC 16 amounted to $10.3\,\mathrm{J\,g^{-1}}$, whereas $15.4\,\mathrm{J\,g^{-1}}$ were measured in BC 30 [75]. This tendency is in line with the melting enthalpy of the P3HT-macroinitiators (ΔH_m macro-P3HT 9 $= 13.1\,\mathrm{J\,g^{-1}}$, ΔH_m macro-P3HT 17 $= 16.9\,\mathrm{J\,g^{-1}}$).

3.3 Optical Properties

A common feature of P3HT homopolymer absorption is the characteristic shoulder at 610 nm, arising from P3HT aggregate formation [77]. The absorption of PPerAcr is only weak at this wavelength. Figure 12a shows the absorption curves of the two homopolymers P3HT and PPerAcr, and the block copolymers BC16–25. Clearly, the block copolymer absorption patterns are convoluted spectra of the two homopolymers. The characteristic shoulder of P3HT is also observed in all block copolymer films, with the intensity at 610 nm roughly correlating to the P3HT

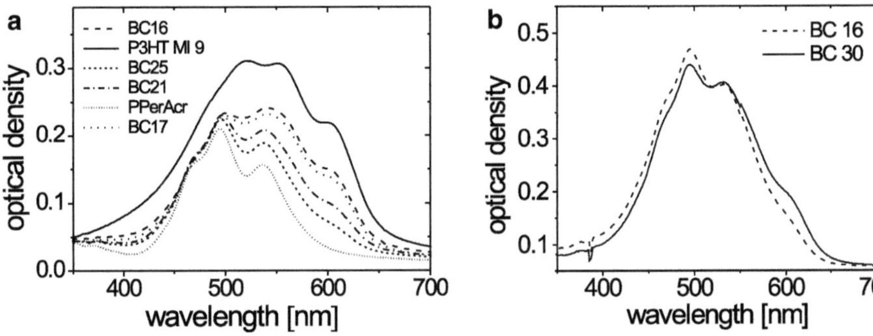

Fig. 12 **a** Optical densities of thin films of P3HT, PPerAcr, and BC 16–25 after spin coating from chloroform followed by chloroform vapour annealing. **b** Optical densities of BC 16 and BC 30 after spin coating from chlorobenzene. (Reproduced with permission from [75]; Copyright 2009 American Institute of Physics)

weight fraction. However, such absorption curves are only observed when the as spun films are subjected to chloroform vapour annealing, since this type of treatment breaks up the π–π interactions between adjacent PBI units in PPerPAcr [78], and therefore gives rise to a rearrangement of the P3HT segments. The P3HT shoulder is only weakly developed in the block copolymer films when using chloroform solutions for spin coating without additional chloroform vapour annealing, since no time for chain rearrangement is given and the films dry out immediately (not shown here).

However, when using high boiling point solvents such as chlorobenzene for spin coating, the development of P3HT aggregates can be observed right after processing. Figure 12b shows the absorption of BC 16 and BC 30 after spin coating from chlorobenzene. Obviously, the optical density at 610 nm is higher for BC 30 than for BC 16. Only recently, a linear correlation between the oscillator strength at 610 nm and the degree of P3HT crystallinity was established [79], suggesting a higher P3HT crystallinity in the film made of BC 30. This result is corroborated by the difference in the melting enthalpies obtained from the DSC measurements (see Sect. 3.2) [75]. The different degrees of P3HT crystallinity of BC 16 and BC 30 having the same composition make these two block copolymers ideal candidates for a comparative study in terms of their morphological and photovoltaic properties, as is attempted in the following.

3.4 *Morphology and Device Performance*

Morphological investigations were carried out using scanning electron microscopy (SEM) on bulk and thin film samples of BC 16 and BC 30 (Fig. 13).

Since thermal annealing of bulk samples did not produce discernible features at the surface, we applied controlled solvent vapour annealing [78] (SVA). This type

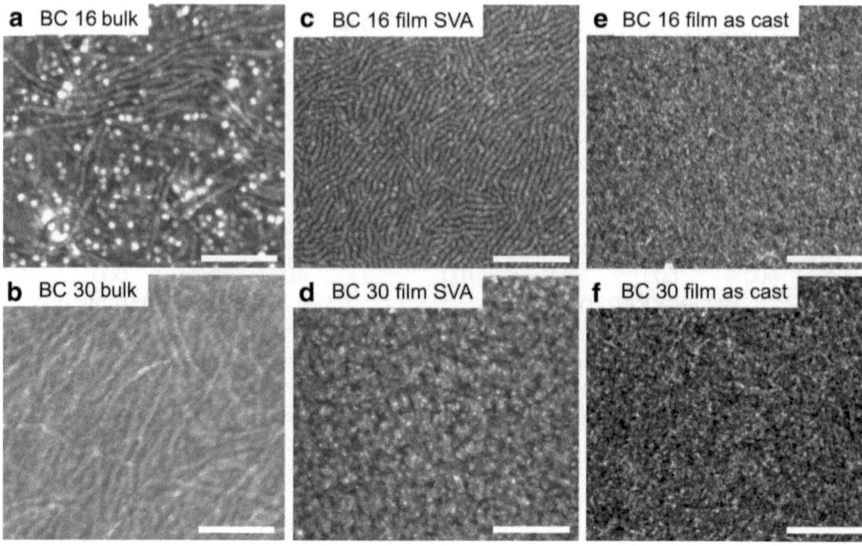

Fig. 13 SEM images of BC 16 and BC 30. **a,b** Surface of bulk samples after solvent vapour annealing (SVA). **c,d** Thin film morphologies after spin coating from chloroform followed by SVA. **e,f** Thin film morphologies after spin coating from chlorobenzene. *Bright domains* are ascribed to PPerAcr and *dark domains* to P3HT. *Scale bar* is 200 nm in all pictures

of treatment led to the formation of fibrilar morphologies in both block copolymers. Figure 13a,b presents the surface of micron-thick bulk samples after SVA. Bright PPerAcr fibrils were observed in a darker matrix of P3HT. While the distance between the fibrils was ~20 nm in BC 16, an increased spacing of ~30 nm was found in BC 30, reflecting the increased M_n of P3HT. A fibrillar structure could also be produced in thin films of BC 16 spun cast from chloroform and subjected to SVA (Fig. 13c). By contrast, thin films of BC 30 were almost featureless at the surface after SVA, even after prolonged annealing times (Fig. 13d). We also tested whether fibrillar morphologies formed after spin coating the block copolymers from higher boiling point solvents such as chlorobenzene. Here, no distinct features were observed in films of both BC 16, and BC 30 (Fig. 13e,f), even though the film of BC 30 suggested slightly larger domains as a result of the higher M_n of P3HT. DSC measurements indicated further evidence for the latter (higher melting point of P3HT block in BC 30).

Photovoltaic devices were produced using the common device architecture indium tin-oxide (ITO)/poly(3,4-ethylenedioxythiophene):poly(styrenesulfonate) (PEDOT:PSS)/block copolymer/aluminium. The solar cell performance was elucidated by measuring the EQE and the current density–voltage (J–V) characteristics (Fig. 14).

Although PPerAcr fibrils formed after SVA in the device of BC 16, the best performance was obtained after simple processing of both block copolymers from chlorobenzene. A possible reason for the lower performance of devices subjected

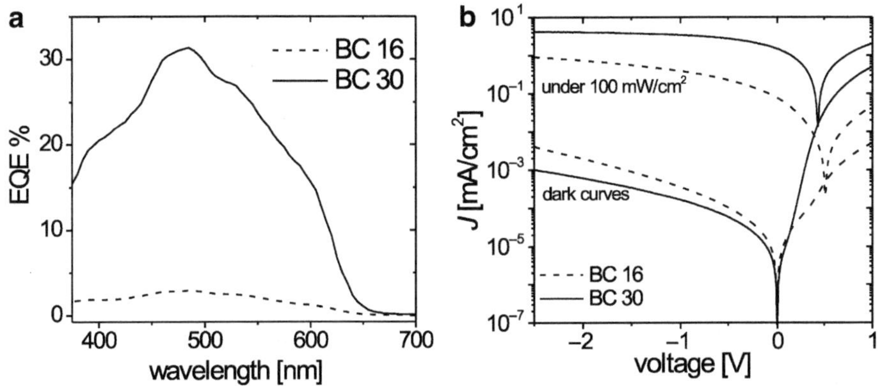

Fig. 14 Photovoltaic performance of BC 16 and BC 30 after processing from chlorobenzene solutions. **a** External quantum efficiency (EQE) and **b** *J–V*-curves under AM 1.5G conditions ($100\,\mathrm{mW\,cm^{-2}}$). (Reproduced with permission from [75]. Copyright 2009 American Institute of Physics)

to SVA might be dewetting, as was observed for a similar block copolymer [80]. Devices with 60 nm thin films gave an outstanding EQE value of 31% for BC 30, whereas BC 16 only delivered 3% (Fig. 14a) [75]. The *J–V* measurements of the device with BC 30 yielded a short circuit current J_{SC} of $1.5\,\mathrm{mA\,cm^{-2}}$, an open circuit voltage V_{OC} of 0.44 V and a PCE η of 0.2% (Fig. 14b). The device with BC 16 as the active layer exhibited $J_{SC} = 0.08\,\mathrm{mA\,cm^{-2}}, V_{OC} = 0.51V$, $\eta = 0.007\%$ and FF = 0.17. We assume the higher degree of crystallinity of P3HT in BC 30 to be the main reason for the drastic improvement of the performance of BC 30 compared to BC 16 (see Sects. 3.2 and 3.3). Also, larger P3HT domains can be expected in films of BC 30, which should result in a better charge percolation. This is obvious from the difference in the molecular weights of the two P3HT macroinitiators employed, but also from the difference of the overall molecular weights (see Table 1). Consequently, the hole carrier mobility is expected to be higher in BC 30 than in BC 16. Indeed, OFET measurements of films of BC 16 and BC 30 revealed hole carrier mobilities of 3×10^{-5} and $5 \times 10^{-3}\,\mathrm{cm^2\,V^{-1}\,s^{-1}}$, respectively [75]. While the huge difference in the hole mobilities of BC 16 and BC 30 is ascribed to P3HT domain percolation and crystallinity here, we note that a dependence of hole transport mobility on molecular weight has already been observed in pristine P3HT homopolymers [81]. However, the dramatic improvement in the hole carrier mobility of BC 30 by more than two orders of magnitude obviously plays a role at the enhanced device performance of BC 30. The high EQE of 31% expected to yield higher PCEs; however, the low fill factor of 0.25 apparently limited the overall device performance. Such photovoltaic parameters appear low when compared to state-of-the-art solar cells [3–5, 7–10]; however, an EQE value of 31% is especially encouraging when considering maximum values of only ~20% obtained for blend devices comprised of P3HT and low molecular weight PBIs [20, 21, 82].

While no EQE measurements were reported in a recent study by Emrick et al. using a similar P3HT-PBI block copolymer, a good PCE of 0.49% was demonstrated [80]. Here, a low molecular weight P3HT block ($5.5\,\mathrm{kg\,mol^{-1}}$) was used, and the devices were thermally annealed prior to electrode deposition. We achieved the best PCE (0.2%) from devices made of BC 30 (M_n of P3HT: $17\,\mathrm{kg\,mol^{-1}}$) after spin casting from chlorobenzene, and thermal annealing led to a decrease of all photovoltaic parameters. Considering the relatively high P3HT chain length in BC 30, we note that the differences in molecular weight, composition and device performance indicate a large potential for further device optimization.

The overall device efficiencies presented here are still moderate compared to conjugated polymer/fullerene based systems [3–5, 7–10]. In this respect we like to point out that by using fully functionalized block copolymers the "idealized morphology" as proposed in Scheme 1 has only been achieved in bulk samples [51, 66], and in thin films with domain orientations parallel to the substrate [78]. We therefore estimate that – among the synthesis of new materials – the largest potential lies within the achievement of vertically oriented block copolymer domains exhibiting a high degree of long-range order in film.

4 Conclusions

Different polymerization methods have been employed to generate donor–acceptor block copolymers with different architectures carrying all functionalities required for the application in OPV cells. Donor–acceptor block copolymers with amorphous poly(triarylamine) segments and side-chain crystalline PBI segments show exciting nanostructures and outperform their analogous blend systems by far. However, the limited charge carrier mobilities of the donor blocks and the moderate absorption are inherent drawbacks of such systems. Relief might be produced by replacing the amorphous donor segment with conjugated, semicrystalline blocks such as P3HT. In this respect, block copolymers comprised of P3HT and side-chain crystalline PBI blocks are promising materials which are just beginning to be investigated. These materials not only address the issues of an improved hole carrier mobility and an extended absorption. More importantly, they enable to confine PBI crystallization to microdomains, thus preventing the uncontrolled growth of crystals observed in blends of conjugated polymers and low molecular weight PBIs. Double-crystalline block copolymers P3HT-*b*-PPerAcr show large differences in device performance when the molecular weight or the composition is varied. On the one hand, these differences point out that structure formation is governed by several processes and therefore complex. One the other hand, these results demonstrate a large potential of device optimization. Further efforts are necessary to fully understand and control the crystallization behaviour and the resulting microstructures, which ultimately determine charge separation, charge recombination, and charge transport.

Acknowledgements Financial support of this work by the Deutsche Forschungsgemeinschaft (SFB 481) is gratefully acknowledged. S.H. acknowledges a scholarship of the Bayerische Graduiertenförderung.

References

1. Tang CW (1986) Appl Phys Lett 48:18
2. Gregg BA (2003) J Phys Chem B 107:4688
3. Wienk MM, Turbiez M, Gilot J, Janssen RAJ (2008) Adv Mater 20:2556
4. Lenes M, Wetzelaer G-JAH, Kooistra FB, Veenstra SC, Hummelen JC, Blom PWM (2008) Adv Mater 20:2116
5. Ballantyne AM, Chen L, Nelson J, Bradley DDC, Astuti Y, Maurano A, Shuttle CG, Durrant JR, Heeney M, Duffy W, McCulloch I (2007) Adv Mater 19:4544
6. Yu G, Gao J, Hummelen JC (1995) Science 270:1789
7. Ma W, Yang C, Gong X, Lee K, Heeger AJ (2005) Adv Funct Mater 15:1617
8. Peet J, Kim JY, Coates NE, Ma WL, Moses D, Heeger AJ, Bazan GC (2007) Nat Mater 6:497
9. Li G, Shrotriya V, Huang J, Yao Y, Moriarty T, Emery K, Yang Y (2005) Nat Mater 4:864
10. Kim J-Y, Lee K, Coates NE, Moses D, Nguyen T-Q, Dante M, Heeger AJ (2007) Science 317:222
11. Arias AC, Corcoran N, Banach M, Friend RH, MacKenzie JD, Huck WTS (2002) Appl Phys Lett 80:1695
12. Wei Q, Nishizawa T, Tajima K, Hashimoto K (2008) Adv Mater 20:2211
13. Gayathri SS, Wielopolski M, Pérez EM, Fernández G, Sánchez L, Viruela R, Ortí E, Guldi DM, Martín N (2009) Angew Chem 48:815
14. Schenning APHJ, Herrikhuyzen Jv, Jonkheijm P, Chen Z, Würthner F, Meijer EW (2002) J Am Chem Soc 124:10252
15. Würthner F, Bauer C, Stepanenko V, Yagai S (2008) Adv Mater 20:1695
16. Wicklein A, Gosh S, Sommer M, Würthner F, Thelakkat M (2009) ACS Nano 3:1107
17. Würthner F, Chen Z, Dehm V, Stepanenko V (2006) Chem Commun 1188
18. Balakrishnan K, Datar A, Oitker R, Chen H, Zuo J, Zang L (2005) J Am Chem Soc 127:10496
19. Dittmer JJ, Marseglia EA, Friend RH (2002) Adv Mater 12:1270
20. Shin WS, Jeong H-H, Kim M-K, Jin S-H, Kim M-R, Lee J-K, Lee JW, Gal Y-S (2006) J Mater Chem 16:384
21. Keivanidis PE, Howard IA, Friend RH (2008) Adv Funct Mater 18:3189
22. Nishizawa T, Tajima K, Hashimoto K (2007) J Mater Chem 17:2440
23. Roncali J (2005) Chem Soc Rev 34:483
24. Tan Z, Hou J, He Y, Zhou E, Yang C, Li Y (2007) Macromolecules 40:1868
25. Hoppe H, Sariciftci NS (2006) J Mater Chem 16:45
26. Campbell AR, Hodgkiss JM, Westenhoff S, Howard IA, Marsh RA, McNeill CR, Friend RH, Greenham NC (2008) Nano Lett 8:3942
27. van Duren J, Yang X, Loos J, Bulle-Lieuwma C, Sieval A, Hummelen J, Janssen R (2004) Adv Funct Mater 14:425
28. Peumans P, Uchida S, Forrest SR (2003) Nature 425:158
29. King S, Sommer M, Huettner S, Thelakkat M, Haque SA (2009) J Mater Chem 19:5436
30. Bates FS, Fredrickson GH (1990) Ann Rev Phys Chem 41:525
31. Buxton G, Clarke N (2006) Phys Rev B 74:085207
32. Kim SH, Misner MJ, Russell TP (2004) Adv Mater 16:2119
33. Thurn-Albrecht T, DeRouchey J, Russell TP, Jaeger HM (2000) Macromolecules 33:3250
34. Angelescu DE, Waller JH, Register RA, Chaikin PM (2005) Adv Mater 17:1878
35. Crossland EJW, Nedelcu M, Ducati C, Ludwigs S, Hillmyer MA, Steiner U, Snaith HJ (2009) Nano Lett 9:2813

36. Stalmach U, de Boer B, Videlot C, van Hutten PF, Hadziioannou G (2000) J Am Chem Soc 122:5464
37. van der Veen MH, de Boer B, Stalmach U, van de Wetering KI, Hadziioannou G (2004) Macromolecules 37:3673
38. Barrau S, Heiser T, Richard F, Brochon C, Ngov C, van de Wetering K, Hadziioannou G, Anokhin DV, Ivanov DA (2008) Macromolecules 41:2701
39. Tu G, Li H, Forster M, Heiderhoff R, Balk LJ, Scherf U (2006) Macromolecules 39:4327
40. Sivula K, Ball ZT, Watanabe N, Frechet JMJ (2006) Adv Mater 18:206
41. Sun S-S, Zhang C, Ledbetter A, Choi S, Seo K, Bonner CE, Drees M, Sariciftci NS (2007) Appl Phys Lett 90:043117
42. Prato M (1997) J Mater Chem 7:1097
43. Dittmer JJ, Lazzaroni R, Leclere P, Moretti P, Granstrom M, Petritsch K, Marseglia EA, Friend RH, Bredas JL, Rost H, Holmes AB (2000) Sol En Mater Sol Cells 61:53
44. Foster S, Finlayson CE, Keivanidis PE, Huang Y-S, Hwang I, Friend RH, Otten MBJ, Lu L-P, Schwartz E, Nolte RJM, Rowan AE (2009) Macromolecules 42:2023
45. Langhals H (1995) Heterocycles 40:477
46. Würthner F (2004) Chem Commun 1564
47. Lindner SM, Thelakkat M (2004) Macromolecules 37:8832
48. Hawker C, Bosman AW, Harth E (2001) Chem Rev 101:3661
49. Benoit D, Chaplinski V, Braslau R, Hawker CJ (1999) J Am Chem Soc 121:3904
50. Sommer M, Thelakkat M (2006) Eur Phys J Appl Phys 36:245
51. Sommer M, Lindner SM, Thelakkat M (2007) Adv Funct Mater 17:1493
52. Yamamoto T, Nishiyama M, Koie Y (1998) Tetrahedron Lett 39:2367
53. Li I, Howell BA, Matyjaszewski K, Shigemoto T, Smith PB, Priddy DB (1995) Macromolecules 28:6692
54. Stolka M, Pai DM, Renfer DS, Yanus JF (1983) J Polym Sci A Polym Chem 21:969
55. Abkowitz M, Pai DM (1986) Philos Mag B 53:193
56. Bellmann E, Shaheen SE, Grubbs RH, Marder SR, Kippelen B, Peyghambarian N (1999) Chem Mater 11:399
57. Maria S, Susha AS, Sommer M, Talapin DV, Rogach AL, Thelakkat M (2008) Macromolecules 41:6081
58. Seo ET, Nelson RF, Fritsch JM, Marcoux LS, Leedy DW, Adams RN (1966) J Am Chem Soc 88:3498
59. Hagopian L, Köhler G, Walter RI (1967) J Phys Chem 71:2290
60. Lindner SM, Kaufmann N, Thelakkat M (2007) Org Electr 8:69
61. Seibt J, Dehm V, Würthner F, Engel V (2007) J Chem Phys 126:164308
62. Huettner S, Sommer M, Thelakkat M (2008) Appl Phys Lett 92:093302
63. Huettner S, Sommer M, Thelakkat M (2010) Appl Phys Lett
64. Barea EM, Garcia-Belmonte G, Sommer M, Huettner S, Bolink HJ, Thelakkat M (2009) Thin Solid Films (in print, doi:10.1016/j.tsf.2009.10.003)
65. Sommer M, Huettner S, Wunder S, Thelakkat M (2008) Adv Mater 20:2523
66. Lindner SM, Huettner S, Chiche A, Thelakkat M, Krausch G (2006) Angew Chem Int Ed 45:3364
67. Loo YL, Register RA, Adamson DH (2000) J Polym Sci B Polym Phys 38:2564
68. Liu J, Sheina E, Kowalewski T, McCullough RD (2002) Angew Chem Int Ed 41:329
69. Iovu MC, Jeffries-El M, Sheina E, Cooper JR, McCullough RD (2005) Polymer 46:8582
70. Iovu MC, Craley CR, Jeffries-El M, Krankowski AB, Zhang R, Kowalewski T, McCullough RD (2007) Macromolecules 40:4733
71. Jeffries-El M, Sauve G, McCullough RD (2004) Adv Mater 16:1017
72. Sommer M, Lang A, Thelakkat M (2008) Angew Chem Int Ed 47:7901
73. Liu J, Loewe RS, McCullough RD (1999) Macromolecules 32:5777
74. Schilinsky P, Asawapirom U, Scherf U, Biele M, Brabec CJ (2005) Chem Mater 17:2175
75. Sommer M, Huettner S, Steiner U, Thelakkat M (2009) Appl Phys Lett 95:183308
76. Hugger S, Thomann R, Heinzel T, Thurn-Albrecht T (2004) Colloid Polym Sci 282:932
77. Clark J, Silva C, Friend RH, Spano FC (2007) Phys Rev Lett 98:206406

78. Huettner S, Sommer M, Chiche A, Krausch G, Steiner U, Thelakkat M (2009) Soft Matter 5:4206
79. Zhokhavets U, Erb T, Gobsch G, Al-Ibrahim M, Ambacher O (2006) Chem Phys Lett 418:347
80. Zhang Q, Cirpan A, Russell TP, Emrick T (2009) Macromolecules 42:1079
81. Kline RJ, McGehee MD, Kadnikova EN, Liu J, Fréchet JMJ, Toney MF (2005) Macromolecules 38:3312
82. Rajaram S, Armstrong PB, Kim BJ, Fréchet JMJ (2009) Chem Mater 21:1775

Adv Polym Sci (2010) 228: 155–207
DOI:10.1007/12_2009_45

Published online: 16 March 2010

Recent Advances in the Improvement of Polymer Electret Films

Dominik P. Erhard, Deliani Lovera, Cosima von Salis-Soglio, Reiner Giesa, Volker Altstädt, and Hans-Werner Schmidt

Abstract Polymer electret materials are electrically charged dielectric polymers capable of quasi-permanently retaining their electric field. However, environmental influences such as temperature and humidity reduce their charge stability and restrict applications. Therefore it is of great importance to provide a broad pool of polymer electret materials and to enhance further the charge storage behavior. In this context we report on concepts, measures, and solutions to improve the electret performance of commodity and high performance thermoplastic polymers, which was carried out at the University of Bayreuth in recent years. It is demonstrated that the commodity polymer polypropylene can be manufactured into excellent electret films when certain trisamide additives are incorporated in very low concentrations. Polypropylene can be employed at temperatures up to its continuous service temperature of 70 °C. To achieve higher temperature windows we investigated the commodity blend system of poly(phenylene ether) (PPE) and polystyrene (PS). We demonstrate that especially PPE/PS blend films with a composition of 75/25 exhibit remarkably good charge storage retention during the isothermal surface decay (ITPD) tests at 120 °C. In addition, the commercially available high performance thermoplastic polyetherimide (PEI) resin containing special phosphorus(III) additives shows very good electret properties at elevated temperatures. These properties can be further enhanced by physical aging; resulting in a charge retention after 24 h at 120 °C as high as 95%. The same beneficial effect of physical aging can be used

D.P. Erhard, C. von Salis-Soglio, R. Giesa, and H.-W. Schmidt (✉)
Makromolekulare Chemie I, Bayreuther Institut für Makromolekülforschung (BIMF), and Bayreuther Zentrum für Kolloide und Grenzflächen (BZKG), Universität Bayreuth, Universitätsstr. 30, 95447 Bayreuth, Germany
e-mail: hans-werner.schmidt@uni-bayreuth.de

D. Lovera and V. Altstädt (✉)
Polymer Engineering and Bayreuther Institut für Makromolekülforschung (BIMF), Universität Bayreuth, Universitätsstr. 30, 95447 Bayreuth, Germany
e-mail: volker.altstaedt@uni-bayreuth.de

to advance PPE and PPE/PS blends. Polymer electret materials with such charge storage properties have the potential to be employed in microphones, sensor devices, and electret filters.

Keywords Charge storage properties · Commodity polymers · Corona charging · Electret materials · High performance polymers · Physical aging · Polymer films

Contents

Abbreviations

CST	Continuous service temperature
DMA	Dynamic mechanical analysis
DSC	Differential scanning calorimetry
E''	Loss modulus
FEP	Fluorinated ethylene propylene copolymer
F-PEI	Fluorinated polyetherimide
i-PP	Isotactic PP
ITPD	Isothermal surface potential decay
M_n	Number average molecular weight
M_w	Weight average molecular weight
NMP	*N*-Methyl-pyrrolidone
PDI	Polydispersity index M_w/M_n
PEEK	Poly ether ether ketone

PEI	Polyetherimide
PEI_{PUR}	Ultem®1000 purified by twofold reprecipitation from dichloromethane into ethanol
PES	Polyethersulfone
PFA	Perfluoralkoxy copolymer
PI	Polyimide
PP	Polypropylene
PPE	Polyphenylene ether, i.e., poly(2,6-dimethyl-1,4-phenylene ether)
PPS	Polyphenylene sulfide
PS	Polystyrene
PTFE	Polytetrafluoroethylene
SIBS	Poly(styrene-*b*-isobutylene-*b*-styrene) triblockcopolymer
T	Temperature
$T_{5\%}$	Temperature at which 5% weight loss occurred in TGA
t_{age}	Period of time at which physical aging was performed
T_{age}	Physical aging temperature
TEM	Transmission electron microscopy
T_g	Glass transition temperature
TGA	Thermogravimetric analysis
T_{ITPD}	Temperature at which ITPD was performed
TSC	Thermally stimulated discharge current
V	Surface potential
V_0	Initial surface potential
Δh	Enthalpy relaxation
ε	Dielectric constant
η_{inh}	Inherent viscosity [dL/g]

1 Introduction

A dielectric material is called *electret* if it exhibits a quasi-permanent electric field, which is either caused by trapping of electrostatic charges (homocharge electret) or by macroscopically oriented dipoles (heterocharge electrets), as illustrated in Fig. 1 [1–3]. While dipole orientation is mainly obtained after cooling a polymer melt in an electric field, real charges can be injected into the material by various methods. Amongst others, corona charging is the most common technique because it is fast, easy to perform, and cost-efficient [4].

The charge storage in materials such as waxes, rosins, and sulfur was recognized in the eighteenth century by Gray. However, the name "electret" was introduced by Heaviside in 1885 since electrets were considered the electrical counterpart of magnets [5]. First systematic investigations of electret materials were performed in 1919 by Eguchi and later by Gemant and Gross [6–8]. Until then, only natural materials had been recognized as electrets. The development and application of electrets advanced substantially along with the discovery that some synthetic polymers are good

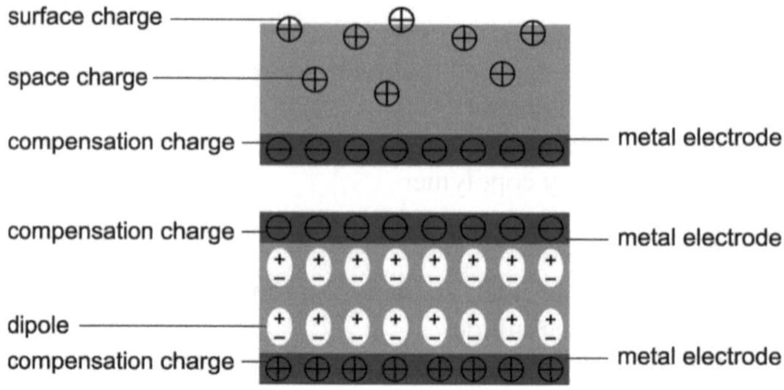

Fig. 1 Schematic of a cross-section of a real-charge electret film with one electrode (*top*) and a dipolar electret film with two electrodes (*bottom*) [1]

electret materials. For example, poly(tetrafluoroethylene) (PTFE) and thermoplastic perfluorinated vinyl ethers (PFA) and fluorinated ethylene propylene copolymers (FEP) exhibit very good charge storage capabilities combined with excellent thermal stabilities. Polypropylene (PP) is also considered a good electret material but additionally an inexpensive commodity polymer and easy to process. However, the relatively low continuous service temperature (CST) [9] of PP represents a limitation for applications in devices operated at higher temperatures.

Electret materials are meanwhile used in a large number of modern high-tech applications including microphones, acoustic sensors, transducers, radiation and pollution dosimeters, power generators, filters, and many more. Additionally, electret technology is of great interest in the field of biomaterials, for instance in callus formation and wound healing [10, 11]. When used in cellular or in multilayer sandwich structures, polymer electrets can exhibit piezoelectricity. Such materials are ferroelectrets, as they show typical features of ferroelectric materials such as piezo- and pyroelectricity [12–17].

The charge storage capability of a polymer electret material is to a large extent determined by the chemical nature of the polymer. In semi-crystalline polyolefins, for example, the crystal domains are embedded in an amorphous matrix and form a heterogeneous solid state morphology, where charge trapping and transport mechanisms depend on the degree of crystallinity. It has been proposed that shallow traps are located in the peripheral and boundary parts of the crystals, while charges with higher release energy are trapped in the central region of the spherulites [18–20]. In amorphous polymers, charge trapping occurs in particular sites of the polymer backbone, for instance at dipoles formed by electronegative atoms, or in between interacting segments of adjacent polymer chains [21–24].

The electret behavior of a polymer material is not only the result of the chemical structure of the polymer itself. Especially when employing commercial resins and films, end groups, catalyst residues, concentration, the type of processing aids and additives, processing conditions, and the thermal history have a great impact on the electret behavior of the particular material. All these parameters are commonly not entirely disclosed by the manufacturers. As will be demonstrated in this chapter, these parameters in particular change the electret performance dramatically. Conversely, the knowledge of these effects can be utilized to tailor and optimize polymer electret materials.

1.1 Electret Properties of Commercial Polymers

To get a general insight into the electret performance of commercial materials, first a comprehensive study on polypropylene resins and films of high-performance polymers was conducted. Generally it is impractical to investigate the various mechanisms of charge storage and charge decay in electrets at room temperature. At such a low temperature, surface and space charges, as well as dipoles, remain immobile in a good electret polymer over a period of decades. Therefore the method of *isothermal surface potential decay* (ITPD) has been introduced which presents an accepted and efficient technique to investigate the intrinsically slow charge drift in polymer electret films [1–3]. Here the characterization of the electret materials is performed by measuring the surface potential of charged films which are annealed under isothermal conditions at elevated temperatures, for example at 90 °C. At these temperatures the trapped charges and dipoles may regain freedom of motion and drift in order to restore electrostatic charge neutrality and/or recombine with ions from the air; thus the surface potential decays in a period of hours.

Among the commodity polymers, *isotactic polypropylene* (i-PP) is an important thermoplastic polymer due to its very good mechanical and thermal properties and consequently broad application spectrum. To identify commercial grades of i-PP and i-PP copolymer resins from different manufacturers which might serve as suitable electret materials and as basis for our research, several types were investigated [25]. Figure 2 shows the ITPD curves at 90 °C of three different i-PP films which were obtained by compression molding. Moplen®HP522H is a grade for the production of biaxially oriented polypropylene films and, according to the manufacturer, does not contain any slip or antiblocking agents nor calcium stearate. The surface potential is rapidly reduced by almost 20% within minutes and then remains constant. After 24 h at 90 °C about 77% of the initial surface potential is still present. The second i-PP grade, Borealis®PP HD601CF, is a film resin also without slip, antiblocking additives, or calcium stearate, and is mainly used for the production of non-oriented films in a chill roll process. The ITPD behavior is very similar to the first grade: After 24 h at 90 °C about 73% of the initial surface potential is still present. The third i-PP resin, Borclean®HB300BF from Borealis, is a high quality film resin specially developed for dielectric applications and also contains neither slip, antiblocking, antistatic, nor nucleating additives. The ITPD curve shows a different decay characteristic. The surface potential decreases more slowly but constantly and after 24 h at 90 °C only about 64% of the initial surface charge is maintained. In summary, Moplen®HP522H exhibits the best electret performance

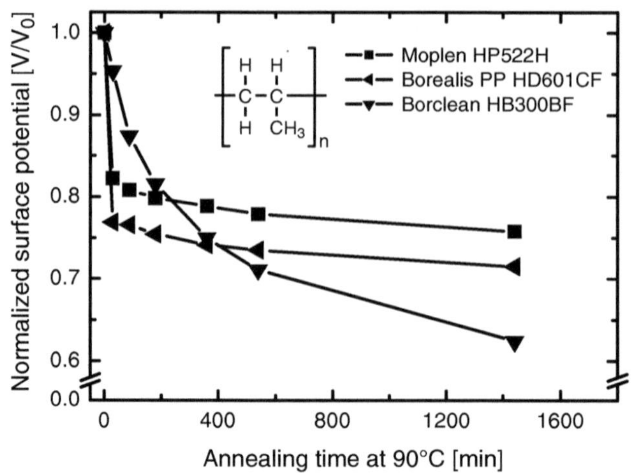

Fig. 2 Isothermal surface potential decay (ITPD) as a function of the annealing time at 90 °C of films (60–80 μm thickness) compression molded using different commercial i-PP homopolymer grades: Moplen®HP522H (Basell), Borealis®PP HD601CF and Borclean®HB300BF (both Borealis)

of all investigated i-PP resins. Other i-PP resins, and in particular i-PP copolymers with ethylene, exhibited worse electret behavior.

On one hand these results show how important it is to select the right grade and manufacturer, but on the other hand indicate that the electret performance of i-PP can be still improved. In Sect. 3 of this review we will present our research on additives which improve the electret performance of i-PP by acting as charge traps.

Generally, commercial polyolefin resins contain a variety of additives, most importantly antioxidants which are typically based on sterically hindered phenols or organophosphite derivatives [26]. Hence we investigated the influence of the primary antioxidant Irganox®1010 (a sterically hindered phenol) and the secondary antioxidant Irgafos®168 (an organophosphite) on the electret properties of i-PP films. To obtain an i-PP sample without additives we used an i-PP reactor powder provided by Borealis, which was additionally extracted successively for 2 days each with acetone, ethanol, and cyclohexane. Afterwards, the above-mentioned antioxidants were re-incorporated into the extracted polypropylene in a concentration range of 0.05–0.3 wt%. The impact on the surface potential decay is shown in Fig. 3. The surface potential decay at different concentrations follows nearly the same trend as the non-additivated reference polypropylene. Only the Irganox®1010 sample with 0.15 wt% shows a small improvement. These results show that the generally applied antioxidants have no influence on the electret performance and consequently commercial grades without extraction were used for further studies.

Of special interest for automotive and filter applications are temperature-stable electret materials which maintain their shape and electret properties at high temperatures, typically around 120 °C. With commodity polymers (Fig. 4) this requirement cannot be fulfilled due to their low CST of max. 100 °C for instance of i-PP.

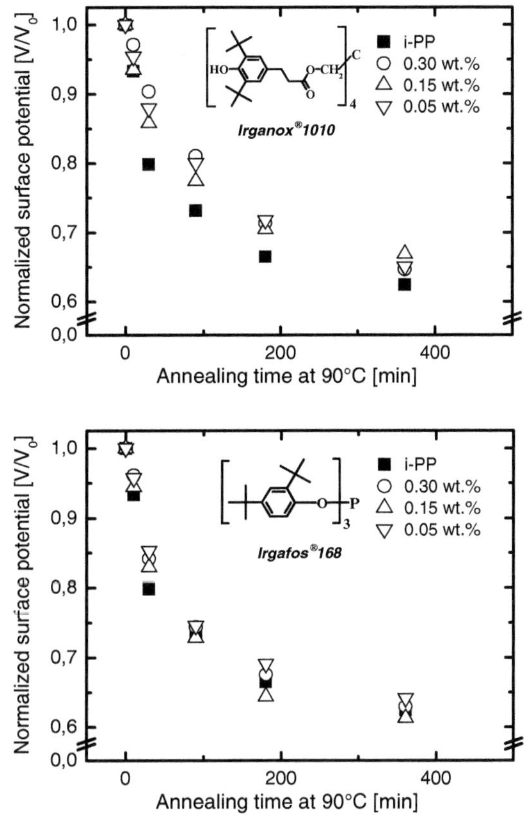

Fig. 3 Surface potential decay curves at 90 °C of i-PP films containing different concentrations of Irganox® 1010 (*top*) and Irgafos® 168 (*bottom*)

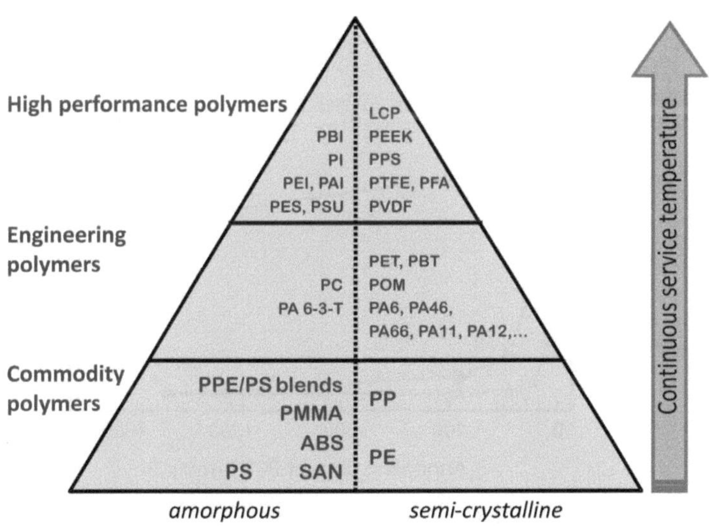

Fig. 4 Commercial polymers classified as commodity polymers, engineering polymers, and high performance polymers. Examples for each class are given in the pyramid

Therefore we investigated several commercial films of *high performance polymers* with respect to their charge storage properties [27]. Among the investigated high performance polymers are poly(ether ether ketone) (PEEK), poly(phenylene sulfide) (PPS), poly(ether sulfone) (PES), polyetherimide (PEI), and polyimides (PI). These high performance polymers exhibit CSTs higher than 160 °C [28], thus exceeding the CST of *engineering polymers*, such as polyethylene terephthalate (PET), polycarbonate (PC), polyamide-6 (PA6), or polyoxymethylene (POM).

Figure 5 displays the ITPD curves determined at 90 °C of PEEK, PES, and PPS films [27]. Both sulfur-containing polymer films exhibit a pronounced charge drain-off within the first 180 min and hence cannot be considered as electret materials. PEEK, on the other hand, shows a moderate surface potential decay and maintains around 50% charge after 24 h at 90 °C. The charge storage characteristics of commercial polyimides and polyetherimide (Ultem® 1000) films at 90 °C are displayed in Fig. 6. Kapton®HN (PI) films lose almost the complete surface potential within 24 h, exhibiting a steep decay in the first minutes. Upilex®R renders better results, since its potential decay is moderate and after 24 h around 45% of the applied charge

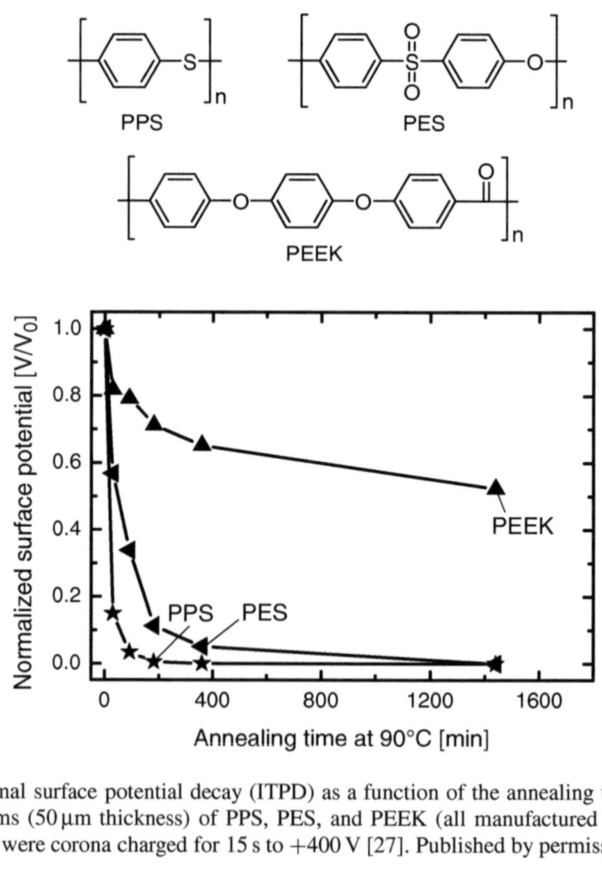

Fig. 5 Isothermal surface potential decay (ITPD) as a function of the annealing time at 90 °C of commercial films (50 μm thickness) of PPS, PES, and PEEK (all manufactured by Lipp-Terler, Austria). Films were corona charged for 15 s to +400 V [27]. Published by permission of WILEY-VCH

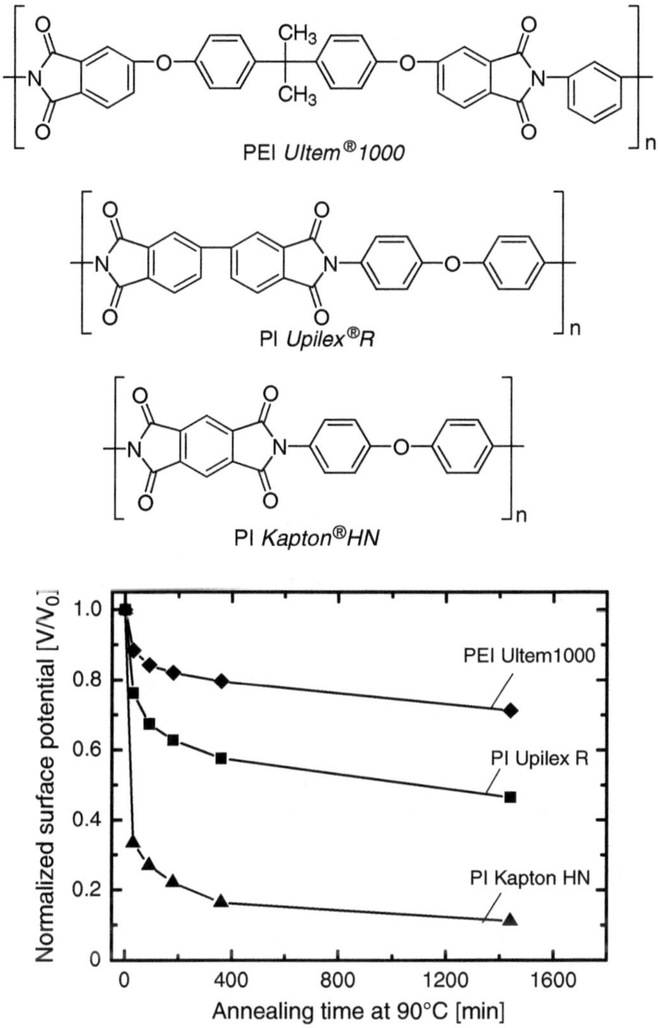

Fig. 6 Surface potential as a function of the annealing time at 90 °C of commercial films (50 μm thickness) of PEI Ultem®1000 (Lipp-Terler), PI Upilex®R (Ube Europe), and PI Kapton®HN (DuPont). Films were corona charged for 15 s to +400 V [27]. Published by permission of WILEY-VCH

is still maintained. The thermoplastic PEI shows even better electret performance. Annealing at 90 °C also leads to charge decay in the first minutes, but the ITPD curve tends to level off at a final charge retention of 70%.

In conclusion, the electret behavior of commercial high performance polymer films depends strongly on the chemical structure and cannot be easily predicted. Among the investigated polymers, PEI films exhibited the best charge storage capability. However, at a remaining surface potential of 70%, further improvement is still desired.

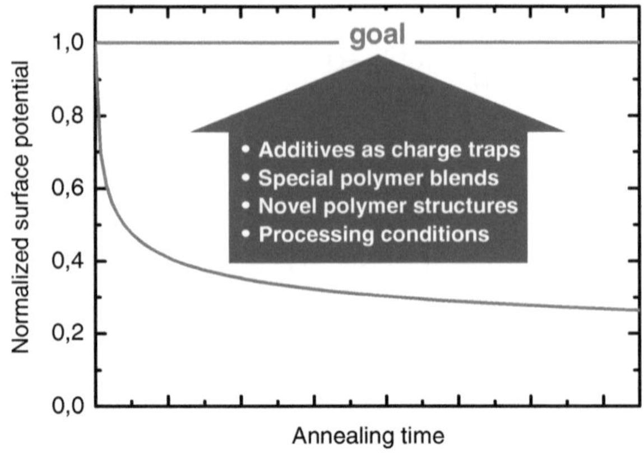

Fig. 7 Schematic illustration of concepts and methods to improve the overall performance of polymer electret materials

Based on these results, our recent research on polymer electret materials focuses on the understanding of (1) the role of the chemical structure of the polymer, (2) the influence of additives, and (3) the variation of processing parameters on the electret performance. As schematically illustrated in Fig. 7, the objective of our research is to provide concepts and methods to improve substantially the electret performance of polymer materials.

This review summarizes our work at the University of Bayreuth over the last few years on improving the electret performance of the *commodity polymer isotactic polypropylene* (Sect. 3) and the *commodity polymer blend system polystyrene/polyphenylene ether* (Sect. 4) to provide electret materials based on inexpensive and easily processable polymers. To open up polymer materials for electret applications at elevated temperatures we concentrated our research on commercially available *high performance thermoplastic polyetherimide* resins and synthesized several *fluorinated polyetherimides* to identify structure-property relations and to improve further the performance at elevated temperatures (Sect. 5).

2 Experimental

Depending on polymer type and final application, various methods are utilized to charge an electret material [1–3]. The most commonly applied method is corona charging since it is easy to perform, fast, cost-efficient, and also allows charging on an industrial scale. Therefore we used mainly corona charging; this method is described in greater detail below. Polymer films with thicknesses between 50 and 100 μm were used as samples. These films were either purchased in the form of sheet films or fabricated by compression molding. In several initial trials we established that all films had a similar thickness and the films underwent the same cooling conditions. Commercial films were used as received.

2.1 Corona Charging

Squares ($40 \times 40\,\mathrm{mm}^2$) of polymer films were mounted onto aluminum plates using conductive double-sided adhesive tape. Samples were charged in a standard point-to-plate corona setup; a photograph of the triode employed is depicted in Fig. 8. The setup was built at the University of Bayreuth with the helpful input of G.M. Sessler and J. Hillenbrand (Technical University of Darmstadt). The schematic circuit diagram of the corona triode is sketched in Fig. 9 [4]. A grid (mesh size 0.2 mm) ensures a homogeneous electric field across the film surface. In preliminary experiments, charging conditions such as voltage, grid size, and distances were optimized. A tip-to-grid distance of 80 mm and a grid-to-specimen distance of 5 mm were employed. The corona voltage was set to +12.5 kV and the grid voltage to +400 V, resulting in uniform film surface potentials between 400 and 420 V. The minimum charging time in the range of 15–30 s was optimized for each polymer.

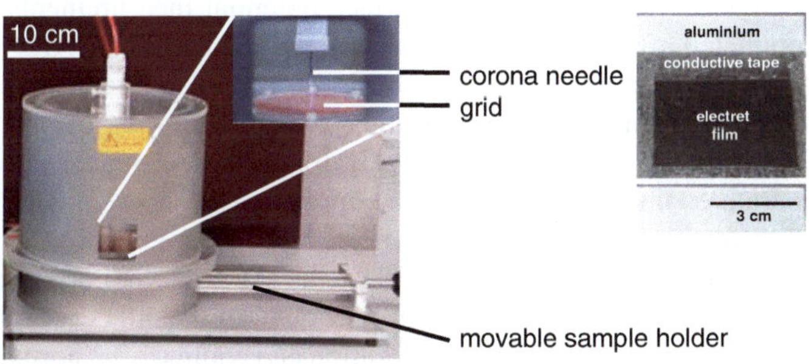

Fig. 8 The *left picture* shows the corona setup. In the *insert* the corona needle and the grid electrode inside the corona chamber are shown. The *picture on the right* shows a polymer film sample mounted on the aluminum electrode with conductive, double-sided adhesive tape

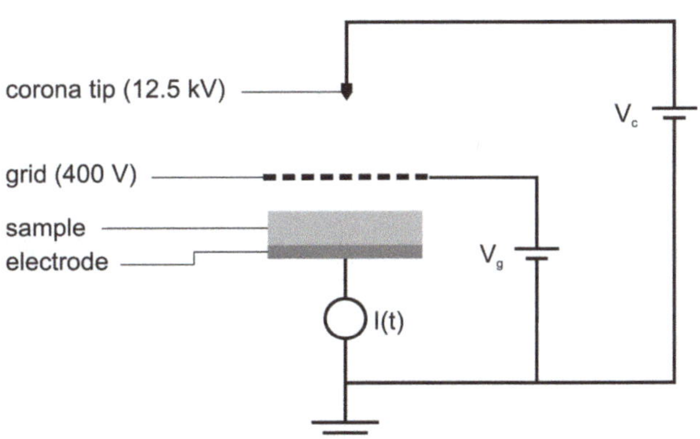

Fig. 9 Schematic circuit diagram of the corona triode setup

2.2 Isothermal Surface Potential Decay

The charge storage characteristics of the corona charged films were evaluated by the isothermal surface potential decay (ITPD) method. Here the surface potential of the electrets films is determined as a function of the exposed time at elevated isothermal temperatures. Under these conditions the trapped charges and dipoles may regain freedom of motion and drift in order to restore electrostatic charge neutrality and/or recombine with ions from the air, causing the surface potential to decay much faster than at room temperature. The electret film samples were placed on top of a computer-controlled x–y-table and the surface potential was measured by a capacitive probe connected to an electrostatic voltmeter (Monroe Electrostatic Voltmeter 244A) at nine defined locations 2 mm above the sample surface on each film (Fig. 10).

The first measurement was taken at room temperature immediately after charging. To accelerate the charge decay, the substrates were then kept at a constant temperature in a convection oven. The surface potential measurements were successively conducted after 30, 90, 180, 360, and 1,440 min (24 h). After each time period the films were removed from the oven, cooled to room temperature, the remaining surface potential measured, and then returned into the oven. For PP the isothermal temperature was 90 °C [27, 29–33]; for high performance polymer films higher temperatures up to 160 °C might be applied. The reported data reflect the average of 3 different films with 9 values (i.e., 27 surface potential values in total); the deviation was below 5%. Figure 11 displays the ITPD curves corresponding to 9 different locations on one i-PP electret film. It is obvious that all curves are within a deviation of less than 5%. For better visualization and comparison, normalized data, meaning the ratio of initial and actual surface potential, are used to plot ITPD decay curves. In the plots shown in this work either the curve based on those averaged values at several time periods is shown or the value of the averaged normalized surface potential value after 24 h is given.

Although several pathways can be held responsible for the charge decay in a film, we believe that the decay is dominated by drift through the bulk and that possible surface conduction mechanisms play a minor role. First, charges migrate relatively quickly into the upper layer of the bulk film and thus away from the surface. Second,

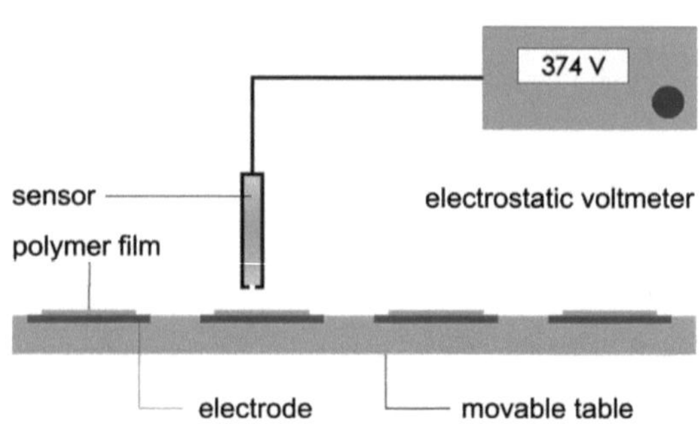

Fig. 10 Schematic drawing of the surface potential measurement setup

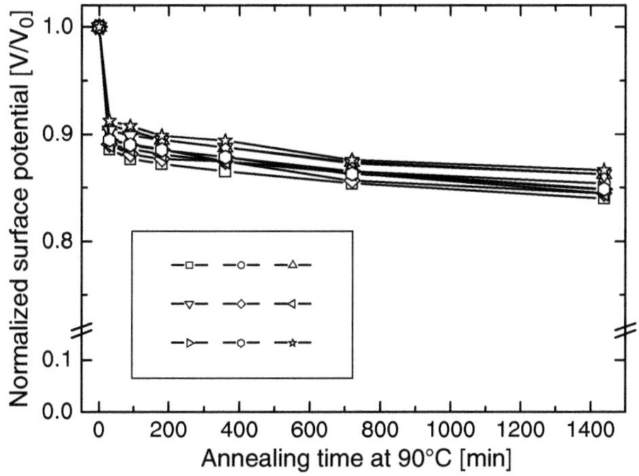

Fig. 11 Isothermal potential decay (ITPD) curves of 9 different locations of an i-PP film sample, distributed in a similar pattern as illustrated. Please note the break in the y-axis

the surface potential is measured strictly around the center of a large film area; hence charge neutralization at the edges is negligible. As a consequence, the surface potential decay at the measurement positions is governed by the charge drift through the bulk.

2.3 *Thermally Stimulated Discharge Current*

The discharge of the polymer electrets can also be characterized by thermally stimulated discharge current (TSC) measurements using an open-circuit setup as depicted in Fig. 12. The charged film is heated linearly at 200 K/h (3.33 K/min) inside an electronically controlled oven (Heraeus 5042EK). A circular induction electrode of 50 mm diameter was mounted 3 mm above the non-metalized surface of the charged film and the discharge current between this electrode and the sample holder was determined by a sensitive electrometer (Keithley 600B). The actual temperature was measured close to the sample with a PT1000 and temperature and discharge current were recorded, resulting in TSC curves similar to the one displayed in Fig. 13. The advantage of this method is that it gives information about the different levels of charge depth and release in the electret film. First, the polarity of the measured current is directly related to the released charge type. A homocharge shows a current of opposite polarity regarding the initial corona charging, whereas a heterocharge exhibits a current of the same polarity. Second, the position and size of the signals is directly related to intrinsic relaxation temperatures, such as T_g, β, or γ relaxations [34].

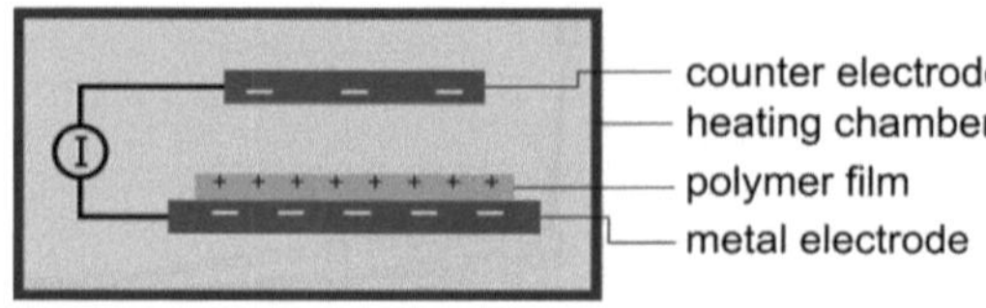

Fig. 12 Schematic drawing of the open-circuit thermally stimulated discharge current setup

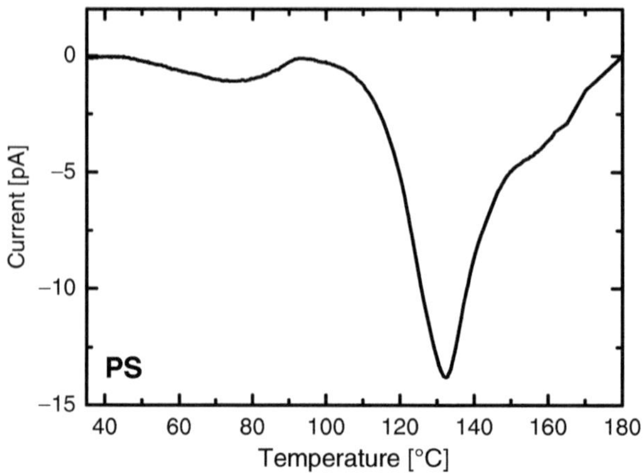

Fig. 13 Representative TSC curve of a 100 μm-thick compression molded polystyrene film. Heating rate was 200 K/h

3 Polypropylene Electrets

It is a challenge and of commercial interest to enhance the charge storage behavior of commodity polymers. The main arguments are simple processability and a low price compared to fluorinated electret materials. Of particular interest in this context is isotactic polypropylene (i-PP). Two principal strategies are applied to improve the electret properties of i-PP. It was demonstrated that a substantial improvement of the charge stability can be achieved by introducing elongated voids during processing, resulting in porous films, although at the expense of reduced transparency and mechanical properties [32, 35–37]. The electret properties of polypropylene films can also be improved by selected additives that act as charge traps as illustrated in Fig. 14. This second approach will be discussed in this chapter in greater detail. Applicable additives must have the ability to form isolated nanostructures at very low concentrations in order to act efficiently as charge traps.

3.1 Trisamides as Electret Additives

Recently, a new class of efficient nucleating/clarifying agents based on 1,3,5-benzenetrisamide chemistry was introduced [38–40]. Due to the stable amide linkage, these compounds possess excellent thermal and chemical stability and

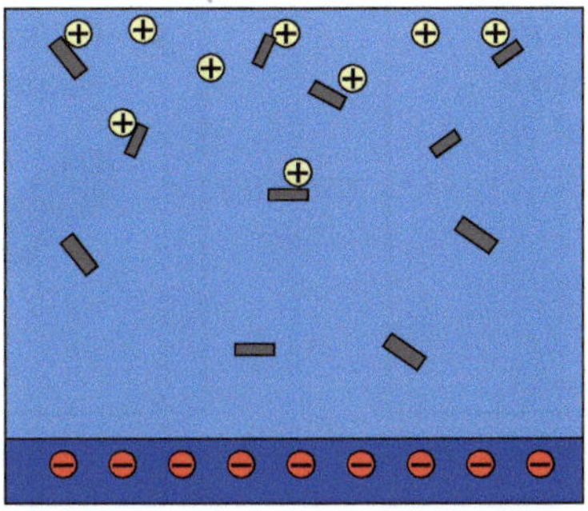

Fig. 14 Schematic representation of an electret film in which charges are trapped by additives

Fig. 15 General structure of trisamides investigated as electret additives

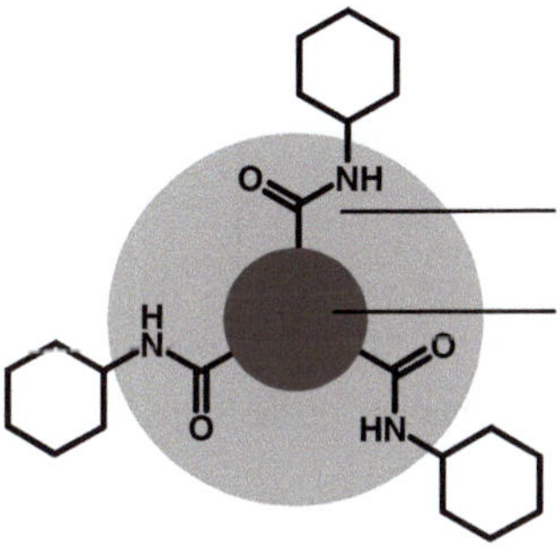

are – depending on their chemical structure, concentration, and temperature – soluble in the polypropylene melt and a few are able to efficiently nucleate and clarify i-PP. The high efficiency is reflected by the very low additive concentration required (below 200 ppm).

Here, we report on the use of trisamides as electret additives in i-PP. A general structure of these compounds is shown in Fig. 15. The molecules have a C3-symmetry and consist of a central core, three units capable of forming hydrogen bonds, and nonpolar peripheral substituents. The central core can be a triphenylamine, benzene, or cyclohexyl unit. The hydrogen bonds are in most cases formed via amide groups and the peripheral substituents consist of alicyclic or aliphatic linear or branched hydrocarbons. The direction of the linkage between the core and the one, two, or all three substituents can be inverted so that either the carbonyl or the amine groups are attached to the core.

Figure 16 illustrates the formation of charge traps upon cooling from the i-PP melt via self-assembly of trisamides. First, the additive needs to be fully dissolved in the melt (A). During cooling, primary supramolecular columns develop due to intermolecular stacking of the molecules and the formation of hydrogen bonds (B). In this self-assembly, process columnar macrodipoles are created which successively self-assemble into supramolecular nano-objects that consist of several columns (C). Upon crystallization of the polymer, these nano-objects stay separated from each other and can act es charge traps (D). It should be pointed out that isolated

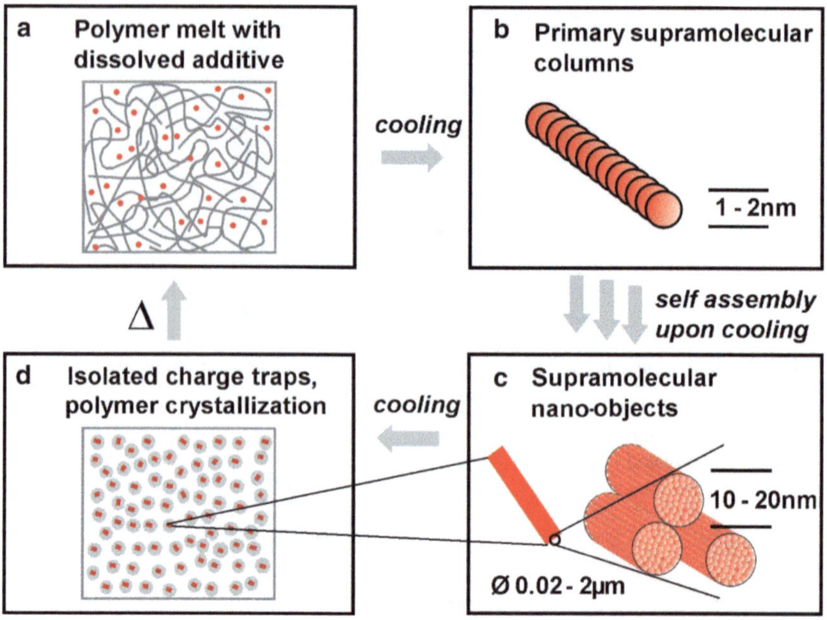

Fig. 16 Schematic representation of the steps to create supramolecular charge traps via self-assembly of trisamides

Scheme 1 Chemical structures of the 1,3,5-benzenetrisamides **1** and **2**

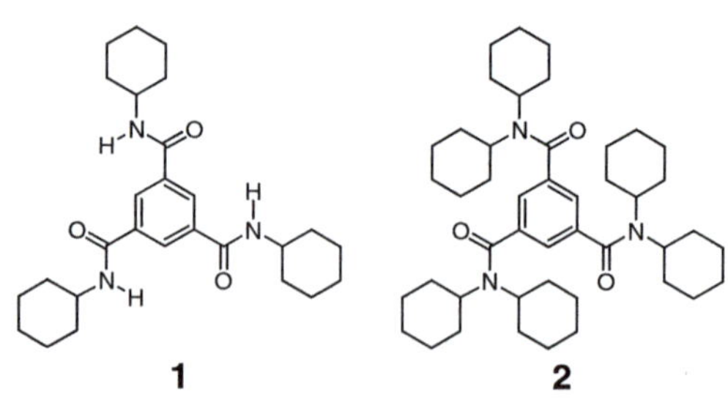

nano-aggregates are formed only in a defined concentration range of the additive. At higher concentrations, percolation and the formation of a three-dimensional network occur, which promotes charge drift.

The following comparison demonstrates the importance of self-assembly and the formation of macrodipoles to yield efficient electret additives. Compounds **1** and **2** (Scheme 1) are trimesic acid derivatives; **1** comprises a trisamide with three secondary amide groups and cyclohexyl substituents, whereas **2** features tertiary amide groups and fully substituted cyclohexyl moieties. As a consequence, compound **2** has no amide protons with which to build intermolecular hydrogen bonds.

In Fig. 17 the ITPD curves of i-PP films at 90 °C, with and without 0.003 wt% of additive **1** and **2**, respectively, are presented. It is obvious that only **1** improves the charge storage behavior of i-PP compared to neat i-PP. The surface potential decay

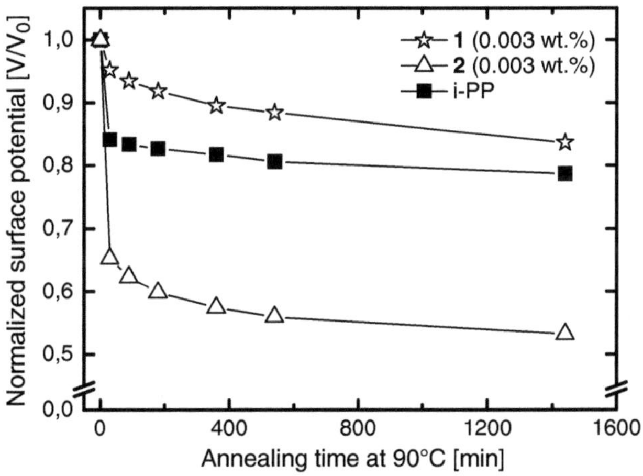

Fig. 17 Isothermal surface potential decay (ITPD) curves at 90 °C for equal concentrations of **1** and **2** in i-PP films and of neat i-PP films. While compound **1** enhances the electret properties of i-PP films, compound **2** causes a fast decay of the charge

curve of i-PP containing **2** quickly drops far below the value of neat i-PP. This clearly demonstrates the difference between both compounds in the performance as electret additive and consequently the requirement of self-assembly and formation of macrodipoles in isolated additive domains.

3.1.1 Additive Concentration Dependence

It was found that a strong correlation exists between the charge storage properties and the concentration of the additive [29, 31]. In order to obtain excellent electret characteristics, the formation of a three-dimensional network structure of the additive, which acts as pathway for charge drift and neutralization, has to be avoided and isolated additive domains are essential. Therefore, the additive concentration has to be decreased to a certain level – depending on the additive solubility and phase behavior – to generate isolated additive domains in the form of supramolecular nano-aggregates. In the following, two examples are given to demonstrate a strong concentration dependence.

The first example is compound **3**, which is a triphenylamine-based trisamide with cyclohexyl substitution. The influence of the additive concentration on the charge storage properties is shown in Fig. 18. Three concentration ranges can be distinguished: range III from 0.25 to 0.15 wt% represents a heterogeneous system where the additive is not completely soluble in the polymer melt at the processing temperature of 260 °C, leading to non-dissolved additive particles. Range II from 0.15 to 0.03 wt% defines the optimum concentration for nucleation. Here, the additive is completely dissolved and, upon cooling, rod-like nanostructures are formed. In this range the surface potential reveals a minimum because the percolating network

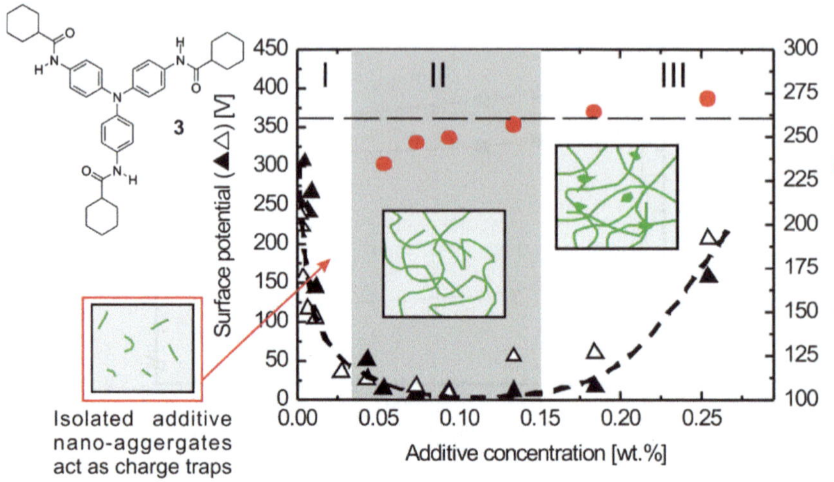

Fig. 18 Dissolution temperature of additive **3** (*red filled circles*) and surface potential of 50 μm-thick i-PP films after annealing at 90 °C for 720 min, as a function of the additive concentration of **3** for two independently prepared series (*filled triangles, open triangles*). The end of the region II (grey) at 0.15 wt% marks the concentration below which the additive is completely soluble in the polypropylene melt at 260 °C (*horizontal dashed line*) [29]. Reprinted by permission of Elsevier Ltd

of the rod-like nanostructures provides pathways which allow the charges to drift. Below concentrations of 0.03 wt% (range I) the surface potential increases considerably, which makes this range very interesting. Here, the rod-like nanostructures are not in contact anymore and thus isolated additive nano-domains arise that act as charge traps [29].

The second example discussed here is based on the cyclohexyl-substituted 1,3,5-benzenetrisamide **1**. The upper graph of Fig. 19 displays the surface potential decay curves of i-PP containing **1** at five different concentrations ranging from 0.05 wt% down to 0.003 wt%. At the highest concentrations of 0.05 wt% and 0.025 wt% the ITPD curves show improved electret behavior in the beginning, although they level off after 24 h to the same value as the i-PP reference. The three lower concentrations of **1** exhibit significantly enhanced charge retention. The optimum concentration is at 0.013 wt%; 90% of the initial charge is still present after 24 h. The bottom graph of Fig. 19 shows the residual surface potential after 24 h at 90 °C plotted as a function of the concentration of **1**. It is obvious that there is an optimum concentration and that higher concentrations neither improve nor deteriorate the electret properties.

Transmission electron microscopy (TEM) studies were carried out to examine size and shape of the isolated additive domains at these low concentrations. In films with a thickness of 50–100 nm the additives in the polymer can be selectively stained with RuO_4. Figure 20 (left) shows the TEM image of an i-PP film comprising 0.01 wt% of compound **1**. Here, an isolated small additive nanostructure at a thickness of 20 nm is visible. No values regarding the length of the additive aggregate can be given, as this length depends on the position of the rod-like struc-

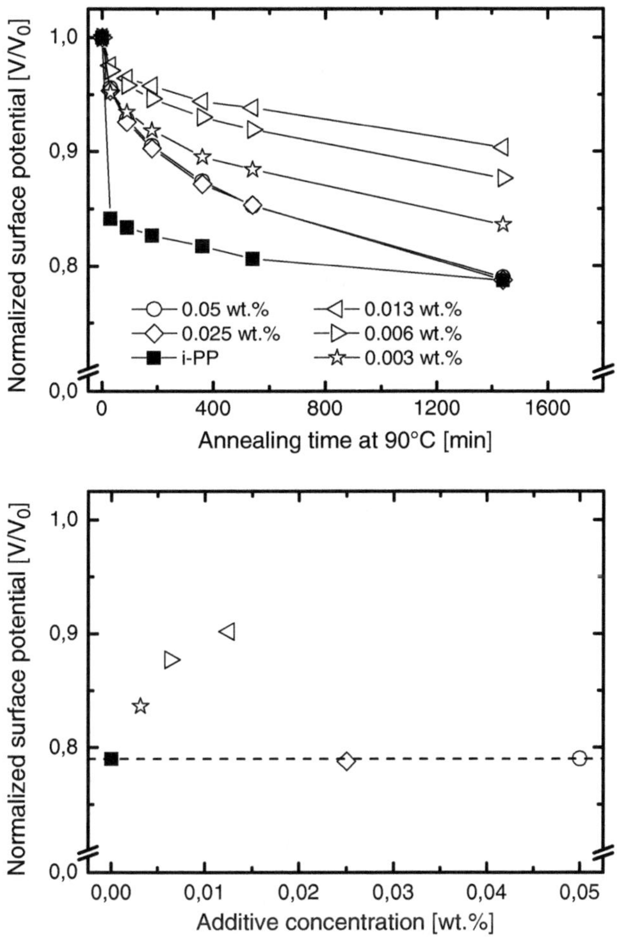

Fig. 19 Surface potential in dependence on annealing time at 90 °C for different concentrations of **1** in i-PP (*top*). The decay curve of neat i-PP (*filled squares*) is included for comparison. Residual surface potential after 24 h at 90 °C as a function of the concentration of additive **1** (*bottom*)

ture with respect to the cutting direction. At a higher concentration of 0.04 wt% of compound **1**, twofold larger additive crystals with a thickness of about 40 nm were found (Fig. 20 right).

3.1.2 Influence of the Cooling Rate

To study cooling rate effects which might affect the supramolecular self-assembly of trisamides in i-PP, we investigated the influence of the processing conditions on the electret performance. Isotactic polypropylene containing 0.09 wt% of compound **1** was chosen for these experiments. After compression molding, the films

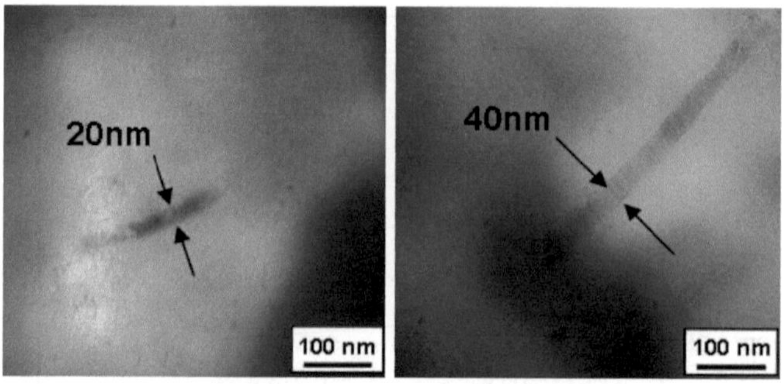

Fig. 20 Transmission electron microscopy images of i-PP films containing compound **1** at 0.01 wt% (*left*) and 0.04 wt% (*right*) showing isolated nano-aggregates of the additive [31]. Reprinted by permission of Elsevier Ltd

were cooled at different rates. At a cooling rate of 10 K/min a three-dimensional fibrillar network structure is formed and a surface potential of only about 50% is retained, which is substantially below the i-PP reference (cooled at the same rate). The surface potential decay curve is shown in Fig. 21 (top) with a polarized optical light microscopy picture of a corresponding sample between crossed polarizers taken at 170 °C immediately after melting the sample. The picture clearly shows the pronounced birefringent network structure of the additive, which is responsible for the rapid discharge. A significant improvement of the charge storage behavior was achieved by utilizing a higher cooling rate of 300 K/min (Fig. 21 bottom). The charge retention after 24 h at 90 °C was 89% of the initial value and higher than that found for the reference material cooled under the same conditions. In this case, optical light microscopy between crossed polarizers revealed no birefringent traces of the additive, as expected.

In addition, the same experiments were carried out at much lower additive concentrations of 0.0024 wt% and 0.005 wt%, corresponding to range I in Fig. 18. In these experiments no improvement of the electret properties by increasing the cooling rate was observed. This can be explained by the fact that in range I, isolated supramolecular aggregates are already formed at the relatively low cooling rate of 10 K/min.

3.1.3 Influence of the Charging Conditions

Two charging methods were applied to study the influence of the charging conditions [41]. The samples were either *surface* charged by a corona method or *volume* charged by mono-energetic electron beams of different energies, achieving penetration depths up to 6 μm. In all cases, surface potentials of about 200 V were chosen. For these investigations we used additive **4** (Scheme 2), a trisamide derivative with cyclohexyl substitution and, compared to compound **1**, one inverted amide linkage. The charge decay in isotactic polypropylene (i-PP) films of 50 μm thickness containing additive **4** at concentrations in the range of 0.004–0.250 wt%

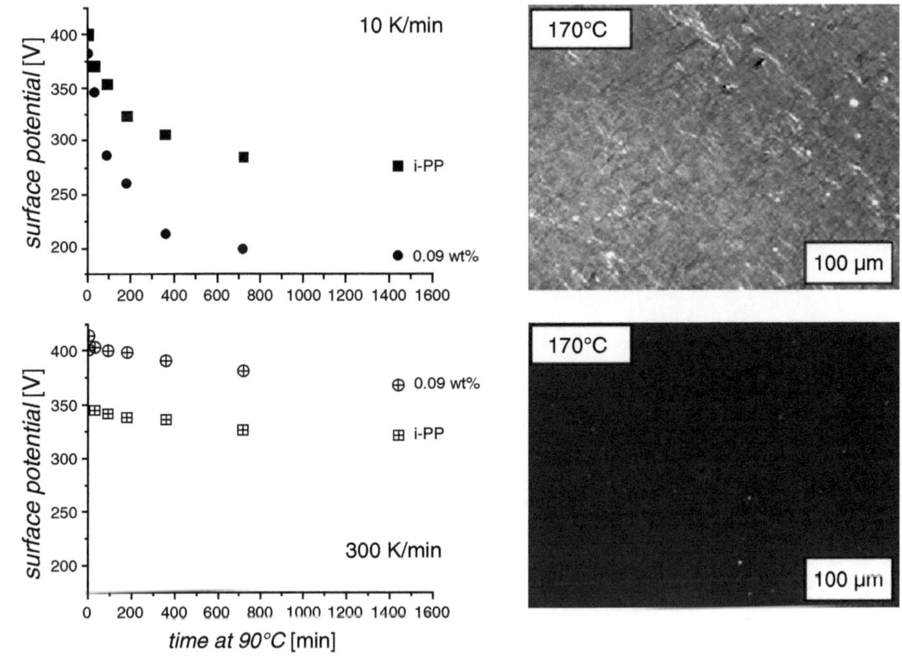

Fig. 21 Surface potential as a function of annealing time at 90 °C of compression-molded i-PP films of 50 µm thickness with compound **1** at a concentration of 0.09 wt% (*circles*) in comparison with the i-PP reference treated under the same conditions (*squares*). The films were produced at different cooling rates of 10 K/min (*top*) and 300 K/min (*bottom*). Corresponding polarized optical microscopy pictures were taken at 170 °C directly after melting the films [31]. Reprinted by permission of Elsevier Ltd

was investigated. After charging the films, the decay of the surface potential was studied by the isothermal discharge method at 90 °C. Results for i-PP films containing 0.125 wt% of additive **4** are shown in Fig. 22. The samples were charged with corona discharge or electron beam at 2, 5, 10, or 15 keV. As far as the effect of the charging method is concerned, no significant difference was found for films with additive **4** between those charged with the corona or electron-beam method.

In Fig. 23 results of TSC measurements are depicted using the same material. The data pertain to corona (top graph) and 10 keV beam-charged samples (bottom graph), respectively. Independent of the charging method, there are peaks at 130–140 °C and at 165 °C, though the 165 °C peak is related to charge release during the melting process of i-PP. The population of the higher-temperature peak, corresponding to the area under the peak, decreases with increasing additive concentration, while the population of the lower-temperature peak increases. A higher-temperature peak means slower isothermal charge decay (larger activation energies), thus lower concentrations of additive **4** are more efficient in charge retention than higher concentrations, as was explained before. A noticeable difference between corona charged and beam charged samples is not observed.

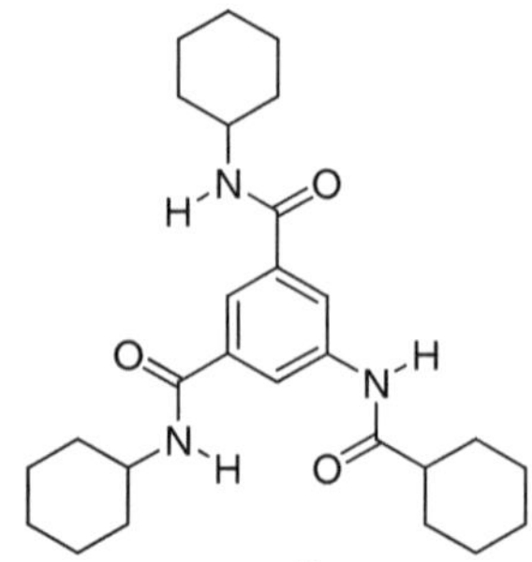

4

Scheme 2 Chemical structure of compound **4**. Note one inverted amide linkage compared to derivative **1**

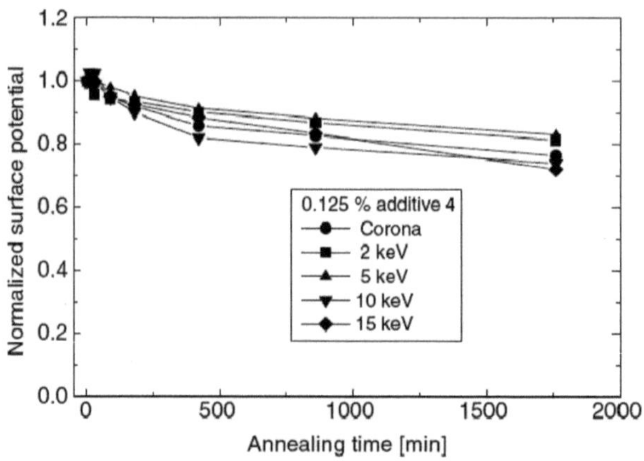

Fig. 22 Isothermal surface potential decay at 90 °C of i-PP films containing 0.125 wt% of additive **4**. The films were corona charged (*filled circles*) or charged with electrons of energies from 2 to 15 keV [41]. Published by permission of IOP Publishing Ltd

3.2 *Bisamides as Electret Additives*

A comparative study on the influence of chemical structure and solubility of a series of low-molecular-weight 1,4-phenylene-bisamides in i-PP was conducted to explore their performance as electret additives [30]. Bisamides are also capable of forming well-defined nano-aggregates via self-assembly from the i-PP melt upon cooling. Here, a comparison of a dicyclohexyl-substituted 1,4-phenylene-bisamide **5** and the asymmetrically substituted cyclohexyl/*n*-alkyl-1,4-phenylenebisamide **6** will be presented; the chemical structures are included in Fig. 24. The symmetry of the molecule, the type of substitution, and the length of the alkyl chain have an influence on the solubility in the i-PP melt while the nucleation efficiency and the ratio of the α- to β-crystal modification will have an effect on the charge storage properties.

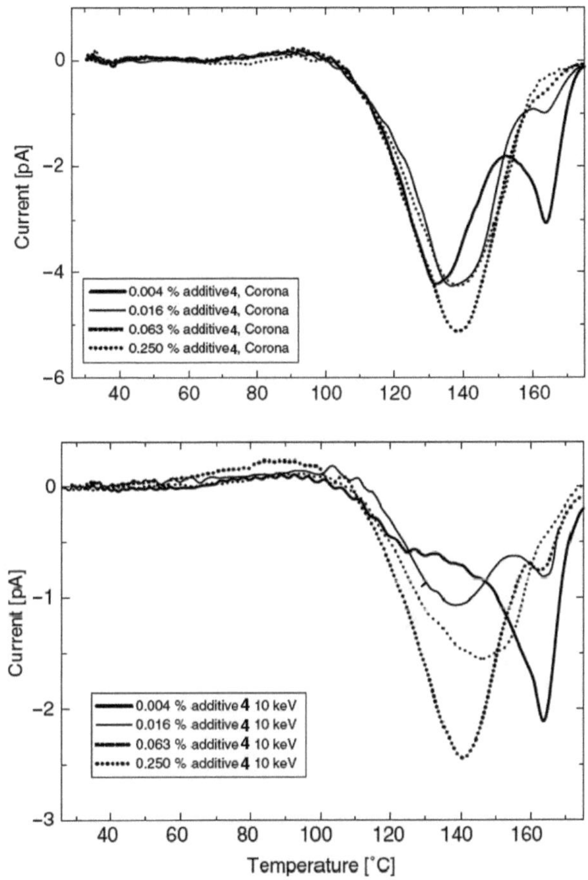

Fig. 23 TSC measurements of corona charged i-PP films containing different concentrations of additive **4** (*top*) and TSC measurements of electron-beam-charged i-PP films with different concentrations of additive **4** (*bottom*) [41]. Published by permission of IOP Publishing Ltd

The additives were incorporated in a concentration range from as low as 0.0005 wt% (5 ppm) up to 0.25 wt% (2,500 ppm). Films containing additive **5** in concentrations up to 0.01 wt. % (100 ppm) exhibit a significantly higher charge storage capability in comparison to neat i-PP, as indicated by a residual surface potential of 360 V after annealing at 90 °C for 24 h (Fig. 24 top). In contrast, the more soluble asymmetric compound **6** with one aliphatic side chain is not capable of improving the charge storage behavior of i-PP over the entire investigated concentration range (Fig. 24 bottom). This observation indicates that such compounds do not form isolated nanostructures, which would act as charge traps.

In this section we have demonstrated that certain nucleating agents based on substituted 1,3,5-benzene trisamides and bisamides are efficient additives to improve

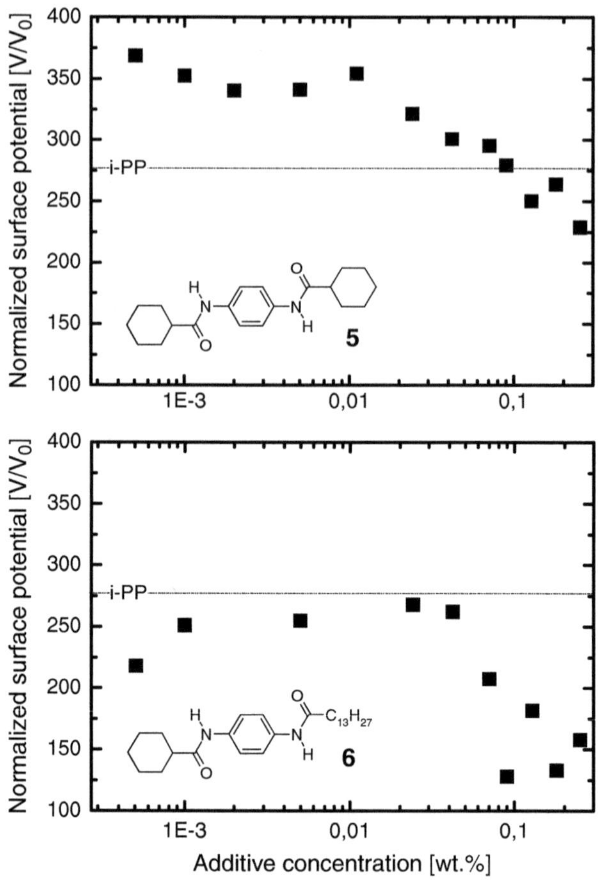

Fig. 24 Additive concentration dependence of the surface potential (after annealing at 90 °C for 24 h) of 50 μm-thick i-PP films comprising bisamide **5** (*top*) and 1,4-phenylene bisamide **6** (*bottom*) [30]. Reprinted by permission of the American Chemical Society

the electret properties of isotactic polypropylene (i-PP). It was found that a strong correlation exists between the charge storage properties and the phase behavior of the additive. In order to obtain excellent electret characteristics, the formation of a three-dimensional network structure of the additive, which acts as a pathway for charge drift and neutralization, has to be prevented and isolated additive domains need to be generated to act as efficient charge traps. We found that, to achieve this, the additive concentration has to be decreased to low levels. Further improvement of the charge storage behavior was achieved by increasing the cooling rate applied during film processing. These results obtained with the selected compounds clearly demonstrate the potential of this class of additives to produce i-PP electret materials.

4 Poly(phenylene Ether)/Polystyrene Blends Electrets

4.1 *Electret Properties of Poly(phenylene Ether) and Polystyrene*

Poly(2,6-dimethyl-1,4-phenylene ether) (PPE), also incorrectly named PPO, is a high-temperature amorphous thermoplastic possessing a glass transition temperature of 215 °C and characterized by relatively low moisture absorption [42], high dielectric strength [43], and excellent mechanical and thermal properties [44]. However, the thermal degradation of PPE starts around its glass transition temperature; hence the neat polymer suffers from degradation during processing, which prevents its commercial use. Blends of PPE and PS, however, can be processed at lower temperatures than neat PPE, which is a successful approach to improve its melt processability. Besides their useful mechanical and thermal properties, PPE/PS blends have been extensively studied as a model system for miscible blends, as they belong to the few polymer pairs which are totally miscible across the entire composition range. PPE is commercially marketed in blends or graft copolymers with polystyrene (PS), for instance as Noryl® or Luranyl®KR resins.

We investigated the electret behavior of PPE and PS films and found that the charge storage performance of PPE is very good [45]. Figure 25 shows the ITPD curves of both blend components at 90 °C. PS films lose their initial charge more rapidly and maintain a surface potential of 58% after 24 h; in contrast, PPE films retain 90% of their initial charge.

The charge storage behavior of neat PPE and PS films was also studied by TSC measurements; the results are plotted in Fig. 26. PS exhibits a small current related to its β-relaxation and discharges rapidly when the temperature exceeds the onset of the α-relaxation (i.e., T_g) due to an increased mobility of charge carriers. The

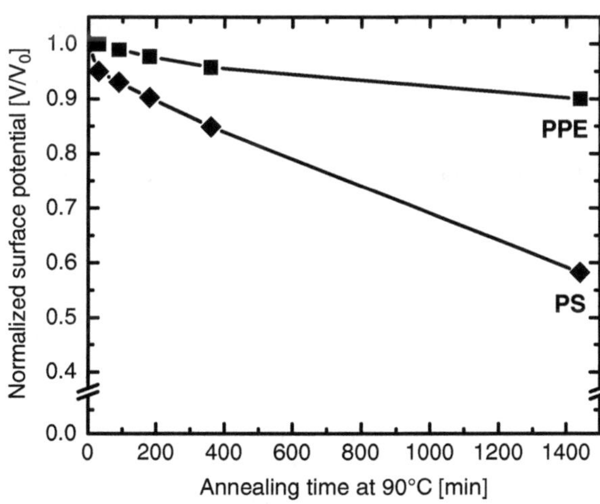

Fig. 25 Isothermal surface potential decay (ITPD) curves of corona charged PPE and PS films (100 µm thickness) as a function of the annealing time at 90 °C [45]. Reprinted by permission of Elsevier Ltd

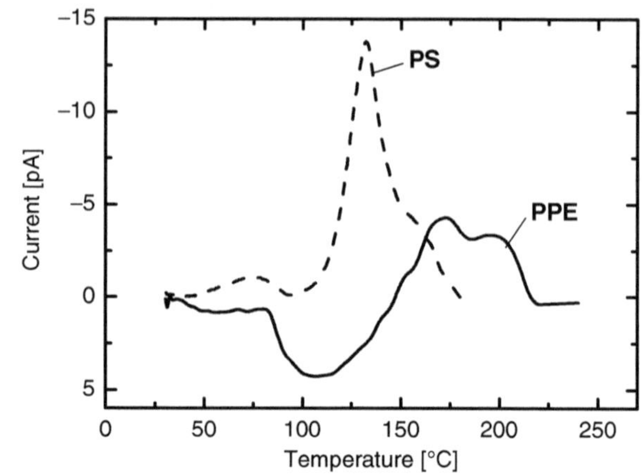

Fig. 26 TSC thermograms of pure PPE and PS films heated at 200 K/h. Films of 100 μm thickness were charged to +400 V for 30 s

TSC thermogram of PPE shows a broad discharge peak between 75 °C and 150 °C. These positive current peaks represent the dipole reorientation in the PPE electret. As the temperature reaches 150 °C, the α-relaxation peak appears in form of two maxima at 166 °C and 188 °C, respectively. This indicates that the charge decay of neat PPE is based on several simultaneously occurring mechanisms. Some polar groups might be aligned in the direction of the electrical field during the charging process [34, 46] or real charges may be trapped at particular sites along the polymer chain, for example at strong electronegative atoms and/or between tightly packed moieties [23].

The reason for the good electret performance of PPE can be explained by the combination of a high glass transition temperature (215 °C) with a low dielectric constant of 2.6 [47]. Insulating polymers with low dielectric constants are generally promising charge storage materials. Indeed, PTFE is the polymer with the lowest dielectric constant ($\varepsilon = 2.10$) and smallest loss of all dielectric materials known so far [48, 49].

4.2 Electret Properties of Poly(phenylene Ether)/Polystyrene Blends

In this section, a systematic evaluation of the charge storage behavior of PPE/PS blends is provided to develop tailored electrets with enhanced temperature stability and processability.

In Fig. 27, the surface potential decay curves of positively charged PPE/PS blend films with three different compositions are presented as a function of the annealing time at 90 °C and compared to neat PS and PPE films. Two blends with 50 wt% PS and 75 wt% PS show a significantly improved electret behavior in comparison to neat PPE. Particularly, the PPE/PS blend 75/25 is capable of maintaining 98%

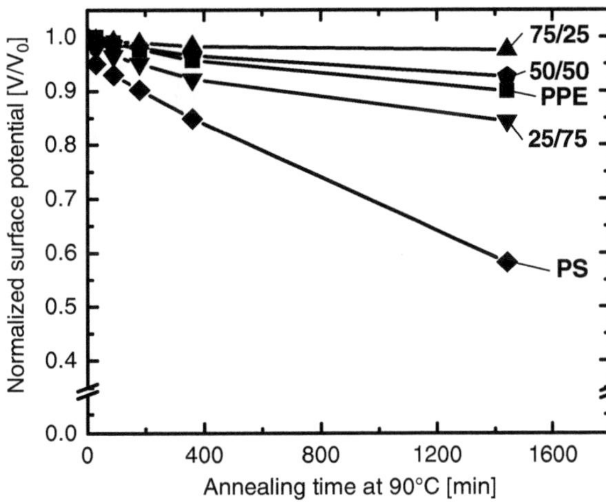

Fig. 27 Isothermal surface potential decay (ITPD) curves of positively corona charged PPE/PS blends as a function of the annealing time at 90 °C. The weight compositions refer to the ratio of PPE/PS [45]. Reprinted by permission of Elsevier Ltd

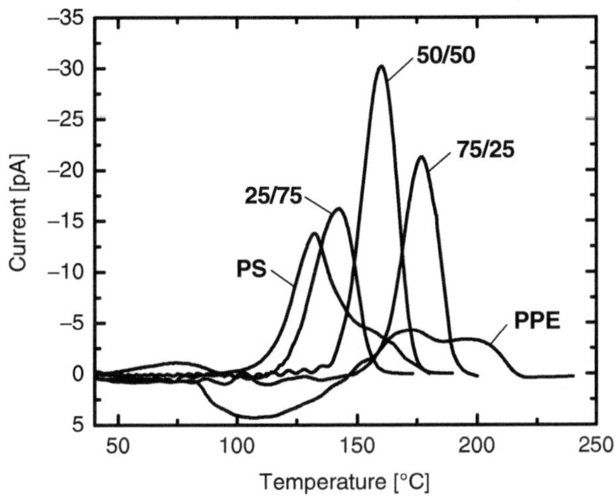

Fig. 28 TSC thermograms of PPE/PS blends compared to neat PPE and PS obtained at a heating rate of 200 K/h. The weight compositions refer to the ratio of PPE/PS. Films of 100 μm thickness were charged to +400 V for 30 s

of its initial surface potential after 24 h of annealing at 90 °C. This experimental observation indicates an unexpectedly strong synergistic behavior and deviates from a simple rule-of-mixture approach.

In Fig. 28, the TSC thermograms of the PPE/PS blend films are again compared to neat PPE and PS films. The maxima of the discharge peaks of the blends were observed at 140 °C for the 25/75 PPE/PS blend, at 160 °C for the 50/50 PPE/PS blend, and at 177 °C for the 75/25 PPE/PS blend. The α peaks in TSC are associated with the glass transition of the blend; thus the release of charges is a consequence of cooperative motions of the chains. Moreover, the blends do not show the low temperature discharge peaks of neat PS around 80 °C.

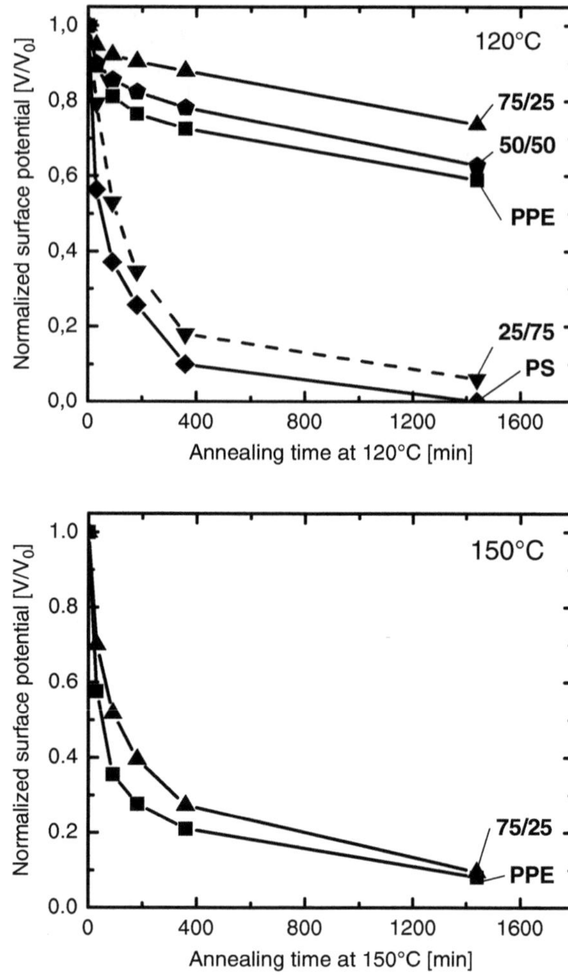

Fig. 29 Isothermal surface potential decay (ITPD) curves of positively corona charged materials as a function of the annealing time at 120 °C (*top*) and 150 °C (*bottom*). The weight compositions refer to the ratio of PPE/PS [45]. Reprinted by permission of Elsevier Ltd

To investigate the potential of these blends for electret applications at elevated temperatures, the annealing temperature for the ITPD experiment was increased to 120 °C and 150 °C; the effect on the surface potential decay is presented in Fig. 29. The complete potential loss at 120 °C of the 25/75 PPE/PS blend and PS films is due to the fact their glass transition temperatures are below this temperature. At 120 °C (Fig. 29 top) the two PPE/PS blends 75/25 and 50/50 are still in the glassy solid state and show similar ITPD trends compared to the curves obtained at 90 °C, but with accelerated decay. The 75/25 PPE/PS blend films still possess a remaining surface potential of 74% after 24 h at 120 °C.

The charge decay of both the neat PPE as well as the 75/25 PPE/PS blend films is noticeably accelerated after raising the temperature to 150 °C (Fig. 29 bottom). After annealing for 24 h, nearly the entire initial surface potential is lost. Although this annealing temperature is still below the onset of the glass transition temperatures

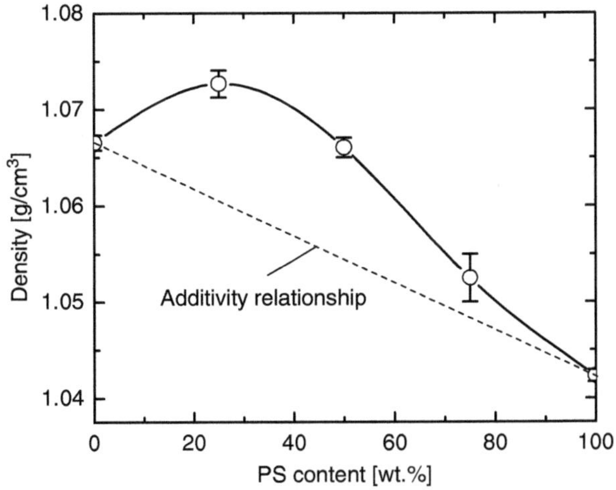

Fig. 30 Bulk film density of the PPE/PS blends as function of the PS weight fraction. The *dashed line* shows a prediction based on a simple rule-of-mixture approach

of both materials at 212 °C and 161 °C, respectively, the relaxation processes and cooperative segmental motions of the polymer chains below T_g are responsible for this accelerated charge decay.

The observed behavior of the PPE/PS blends can be associated with two particular morphological characteristics. The first one is related to the packing density of the blends on a molecular level. A higher packing density of the blends in the glassy state as predicted from a rule-of-mixture approach can indeed be identified by the macroscopic density as a function of the blend composition, as plotted in Fig. 30.

This increase in density, or negative excess volume of mixing, is common for compatible blends with strong intermolecular interactions [50]. Miscibility of PPE and PS is a result of a strong π-π electron interaction between the aromatic rings of PPE and of PS [51, 52]. The maximum in blend density at a PPE weight content of 75 wt% directly correlates to the observed optimum in charge storage behavior. The increased packing density limits segmental motions below T_g and thus effectively reduces the charge drift.

Another peculiarity of the PPE/PS blends on a molecular level also might explain the observed temperature dependence of the surface potential decay behavior. As mentioned before, complete miscibility of PPE/PS across the whole composition range is generally assumed. As shown in Table 1, a detailed DSC study revealed the presence of a single glass transition for all materials. Nevertheless, a negative deviation from the theoretical values predicted by both the Fox [53] and the Couchman equations [54] was calculated. In addition, the ΔT_g values revealed rather broad glass transitions of the blends. These observations indicate the presence of some sort of micro-heterogeneity in the mixed phase and local fluctuations in blend composition [55].

Table 1 Glass transition behavior [T_g and T_g width (ΔT_g)] measured by DSC and DMA as well as theoretical values predicted by Fox [53] and Couchman [54] equations for the neat materials and PPE/PS blends

	DSC			DMA	Calculated	
PPE/PS	$T_{g\text{-ONSET}}$[°C]	$T_{g\text{-MIDPOINT}}$[°C]	ΔT_g[°C]	$T_{g\text{-}E''}$[°C]	$T_{g\text{-FOX}}$[°C]	$T_{g\text{-COUCHMAN}}$[°C]
100/0	212	215	6	213	212	212
75/25	161	169	15	169	178	180
50/50	127	135	15	141	146	148
25/75	106	111	11	116	118	120
0/100	89	94	9	100	89	89

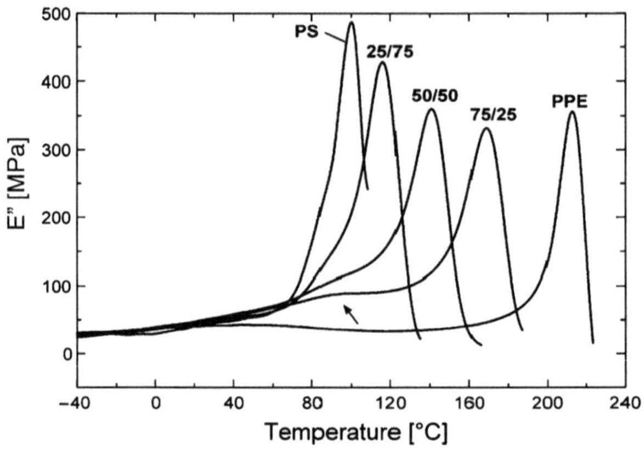

Fig. 31 Loss modulus E'' as a function of the temperature for injection-molded PPE/PS blends as determined by DMA. The low-temperature shoulder at around 90 °C exhibited by the blends is indicated. The weight compositions refer to the ratio of PPE/PS [45]. Reprinted by permission of Elsevier Ltd

To verify additionally the occurrence of such local inhomogeneities, dynamic mechanical analysis (DMA) experiments were conducted on melt-processed specimens. The peak position of the loss modulus (E'') of the blends shown in Fig. 31 reflects the trends observed by DSC. However, a further relaxation phenomenon in the blends at temperatures below the glass transition is revealed by the presence of a low-temperature shoulder at around 90–100 °C, coinciding with the glass transition temperature of PS. This feature is evidence that a certain fraction of PS is not intimately mixed with the PPE. It has previously been reported that PPE/PS blends can show such micro-heterogeneities when the limit of the thermodynamic equilibrium was not reached [56]. PS domains with sizes up to 10 nm have been reported to co-exist in an intimately mixed PPE/PS phase [51, 57]. The resulting nano-scale phase boundaries might also act as charge traps and thus restrict the charge decay of the blends as compared to that of neat PPE and be responsible for the very good electret behavior of the 75/25 PPE/PS blend.

4.3 Physical Aging of Poly(phenylene Ether)/Polystyrene Blends

One of the parameters related to the synergistic behavior observed for the electret performance of PPE/PS blends is the increased packaging density of the blends in comparison to the neat polymers. Amorphous polymers such as PPE and PS undergo physical aging and volume relaxation when they are held at temperatures below their glass transition temperature T_g [58, 59]. The T_g is linked to the α-relaxation, below which the material is rigid and glassy, and above which it is soft, flexible, and rubbery, and is an important parameter for amorphous polymers [60, 61]. The α-relaxation process is generally related to the thermal energy required for changes in the conformation of the polymer backbone and cooperative segmental motions. Below the glass transition, only very slow processes which attempt to establish the equilibrium take place. Annealing of an amorphous polymer at temperatures below its T_g is generally called sub-T_g heat treatment or physical aging [58, 62]. The structural relaxation during physical aging affects various mechanical, optical, and barrier properties [64]. The material becomes stiffer and more brittle, the creep rate decreases, the density increases, and the T_g is shifted to higher temperatures [60, 61]. Since physical aging occurs in a broad temperature range below T_g, amorphous polymer electrets might occasionally undergo physical aging during usage [63].

In this part of the work, a physical aging treatment (Table 2) was employed as a method to increase the density of the blends and to study the effect on charge storage properties. In Fig. 32 the effect of physical aging on the surface potential of PPE/PS blends prior to corona charging is presented as a function of different annealing temperatures T_{ITPD} after 24 h. In addition to the T_{ITPD} of 90 °C, experiments were also conducted at 120 °C and 140 °C. The physical aging conditions had to be adjusted for each blend composition and are summarized in Table 2. Note the difference between the physical aging temperature (T_{age}) and the charge decay acceleration temperature (T_{ITPD}) during the remaining potential measurements.

A beneficial effect of the aging treatment was found for almost all blend compositions and homopolymers. Notable are the very high charge retentions of up to 98% at 90 °C. At a T_{ITPD} of 120 °C, PS and the 25/75 PPE/PS blend films lose all their charge because their T_g is exceeded. The physical aged PPE and the 75/25 PPE/PS blend films performed remarkably well. Charge retention values of 89% and 79%, respectively, were determined. Even at a higher T_{ITPD} of 140 °C, the physically aged

Table 2 Glass transition temperatures T_g, physical aging temperatures prior to corona charging T_{age}, and aging periods t_{age} of the investigated PPE/PS blend compositions in wt%

PPE/PS	T_g [°C]	T_{age} [°C]	t_{age} [h]
100/0	215	160	12
75/25	169	150	12
50/50	135	120	12
25/75	111	95	12
0/100	94	80	12

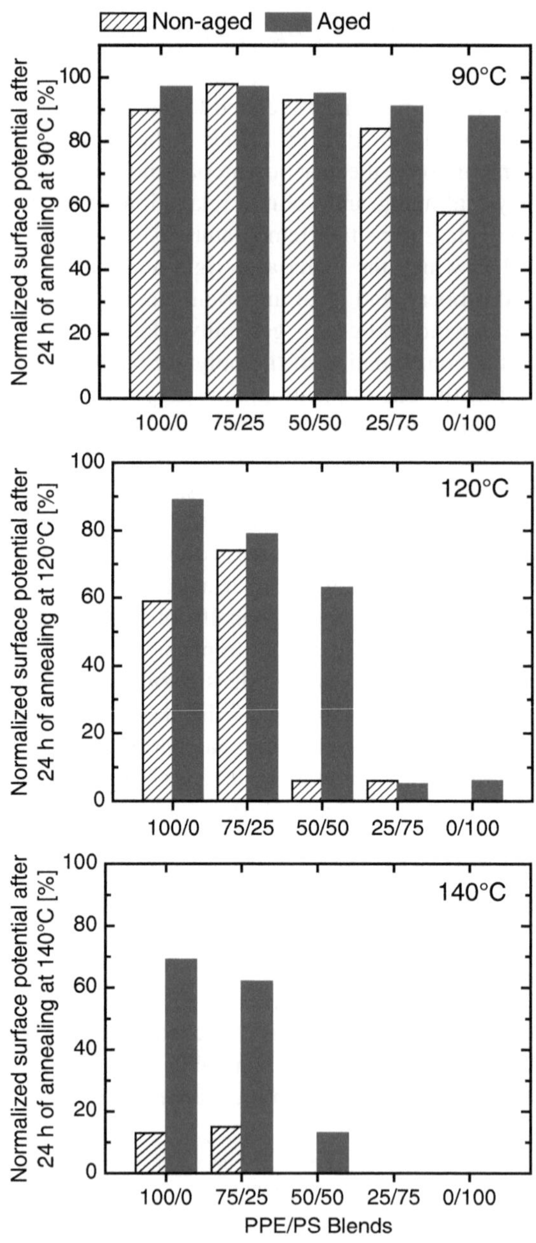

Fig. 32 Surface potential retained by non-aged and physically aged PPE/PS blend films and their corresponding homopolymer films. The aging conditions are presented in Table 2. The ITPD measurements were carried out at $T_{ITPD} = 90\,^{\circ}C$ (*top*), $T_{ITPD} = 120\,^{\circ}C$ (*middle*), and $T_{ITPD} = 140\,^{\circ}C$ (*bottom*) for 24 h

PPE and 75/25 PPE/PS blend films retained 69% and 62% of their initial charge, respectively. These are excellent values considering a T_{ITPD} of 140 °C.

The superior charge retention of the aged PPE compared to the aged PPE/PS blends might be related to the density of the materials prior to physical aging. As

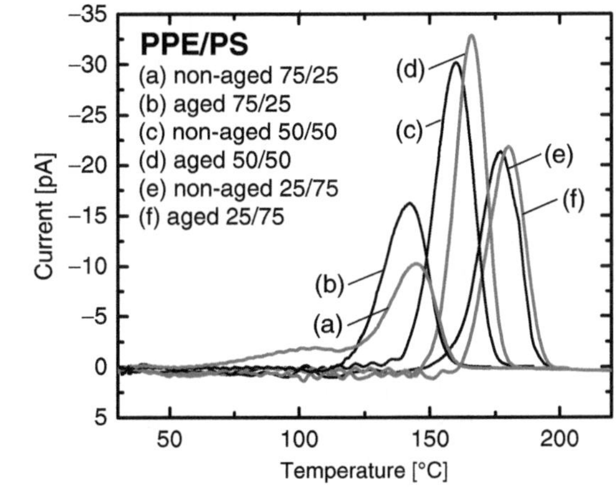

Fig. 33 TSC spectra of PPE/PS blend films 25/75 non-aged (a) and aged (b); PPE/PS blend films 50/50 non-aged (c) and aged (d); PPE/PS blend films 75/25 non-aged (e) and aged (f). Films of 100 μm thickness were charged to +400 V for 30 s

mentioned previously, PPE/PS blends are more densely packed in the glassy state compared to calculated densities based on the pure components. During the aging treatment, the blends gain intrinsically lower mobility, higher relaxation times and slower densification than the rule-of-mixture would suggest. In contrast, PPE has a high fractional free volume which is a driving force in the facilitation of chain packing during physical aging [52]. PPE has a higher volume relaxation rate compared to the blends [64], which means that for the given aging conditions, it would reach a similar degree of packing more quickly than those blends with 50 wt% of PPE and more.

The change of the TSC spectra with the composition of the PPE/PS blends is depicted in Fig. 33. After physical aging, the peaks shift to higher temperatures and show higher currents. However, with increasing amounts of PPE, the aging-induced peaks shift and their amplitudes are diminished. It is clearly visible that, with increasing amounts of PPE, the discharge peak shifts to higher temperatures, indicating an overall higher charge storage stability.

Considering the slower volume relaxation rate that the PPE/PS blends have in comparison to neat PPE, our results show that a physical aging treatment at suitable conditions can effectively reduce the free volume of amorphous materials and lead to further improvement of the charge retention.

4.4 *Charge Storage of Ternary Blends*

As outlined in the previous section, the good charge storage of the binary PPE/PS blends was in part attributed to the proposed presence of heterogeneities in the polymer bulk. In this section, heterogeneities are further introduced by adding a third component, namely the triblockcopolymer poly(styrene-*b*-isobutylene-*b*-styrene) (SIBS) to form ternary polymer blends with PPE and PS [65]. The chemical struc-

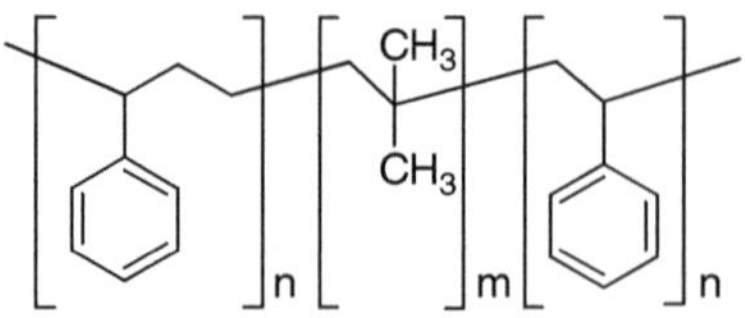

Scheme 3 Chemical structure of the poly(styrene-*b*-isobutylene-*b*-styrene) triblockcopolymer (SIBS) used in this study. The SIBS (grade 103T; Kaneka Co.) had a M_w of 78 kg/mol (PDI = 1.74) at a PS content of 34.2 wt%

ture of the SIBS employed in this study is depicted in Scheme 3. SIBS was chosen since PPE/SIBS blends have been reported in the literature as materials with a good combination of fracture toughness, impact strength, and mechanical properties [66]. In the following, first results on a correlation between electret properties, blend compositions, and final morphologies are presented.

In Fig. 34 the normalized charge decay curves as a function of the annealing time at 90 °C of the PPE/PS 75/25 blends modified with 5, 10, and 20 wt% of SIBS are presented. The ITPD measurements at 90 °C showed no substantial difference regarding different blend compositions. Increasing the ITPD temperature to 120 °C results in a higher mobility of the charge carriers and accelerates the charge decay. Increasing amounts of SIBS increases the charge decay at the beginning. However, the blend with 5 wt% of SIBS possesses a better charge stability than the reference material after 24 h of annealing.

The blend morphology was investigated by transmission electron microscopy. The TEM micrographs of the blend PPE/PS 75/25 modified with SIBS are shown in Fig. 35. The PS domains are readily stained by RuO_4 [67] and appear as the darkest regions in the micrographs, while the miscible PPE/PS matrix is dark grey and the elastomer blocks are imaged as bright regions. In these micrographs, a dispersed "sea-island" blend morphology can be observed, where the copolymer domains have a core-shell morphology formed by grey or non-colored cores with black edges. The elastomeric blocks of SIBS are not miscible in the PPE/PS matrix and hinder the free dispersion of the PS blocks. Therefore, PS segments are only partially dissolved in the matrix and can be distinguished as the dark edges around the dispersed phases or "islands". The dispersed SIBS have a size between 25 and 90 nm in diameter and start to coalesce only when the SIBS content reaches 20 wt%.

Based on these results, it can be concluded that the charge storage of immiscible blends is influenced by the size and degree of dispersion of the dispersed phase. Since the size of the dispersed phase increases with the SIBS content in the blend, a limit of charge storage improvement will exist. In the aforementioned example of the PPE/PS blends with SIBS, such a limit was reached when the dispersed phase reached around 100 nm (5 wt% of SIBS). A larger dispersed phase, as was the case with the PPE/PS blend at 20 wt% SIBS, may decrease the charge storage capability.

In Fig. 36, the surface potential decay of a 75/25 PPE/PS blend film with 2.5 wt% of SIBS was evaluated at three different temperatures (120 °C, 130 °C, and 140 °C) and compared to the corresponding PPE/PS blend. The addition of only 2.5 wt%

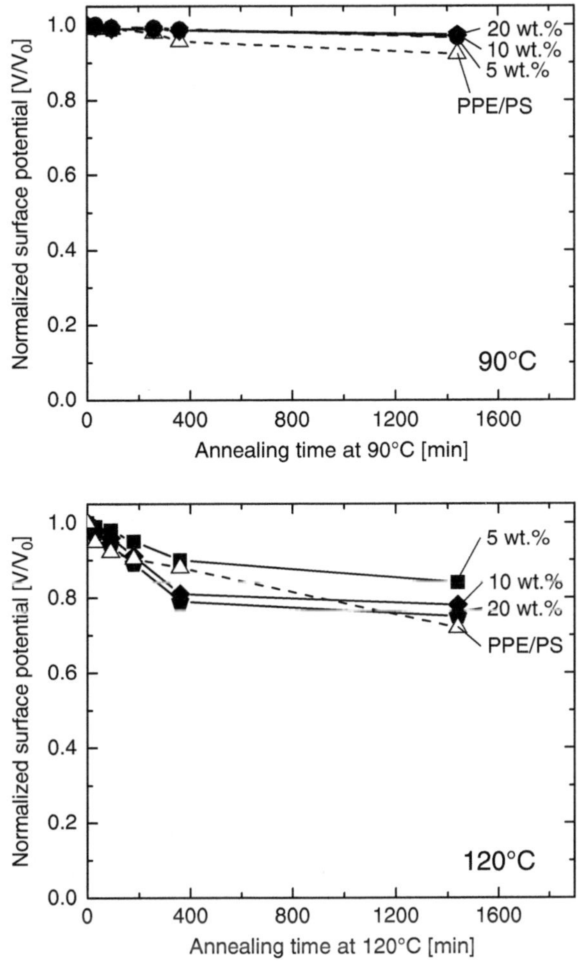

Fig. 34 Isothermal surface potential decay (ITPD) curves as a function of the annealing time of 75/25 PPE/PS blend films modified with 5, 10 and 20 wt% of SIBS at (*top*) 90 °C and (*bottom*) 120 °C [65]. Published by permission of the Society of Chemical Industry

SIBS, which has a particle size between 15 and 70 nm, clearly enhances the charge storage behavior of the 75/25 PPE/PS. After 24 h at $T_{ITPD} = 120\,°C$ the ternary blend exhibits a remaining surface potential of 81% compared to 65% of the corresponding PPE/PS blend. Increasing T_{ITPD} leads to a linear decrease in the retained surface potentials of all films. However, the material obtained with 2.5 wt% SIBS is able to store 59% of the initial charge when annealed at a T_{ITPD} of 130 °C for 24 h.

In summary, this section demonstrates that the intrinsically good electret properties of PPE can be further enhanced by blending with PS. Particularly, the PPE/PS blend with a composition of 75/25 exhibited an excellent charge storage capability, even at elevated temperatures. A similar performance can be achieved with neat PPE in combination with physical aging prior to corona charging. These results support the assumption that the origin of charge decay is related to sub-T_g segmental motions, which are reduced either by blending with PS or by physical

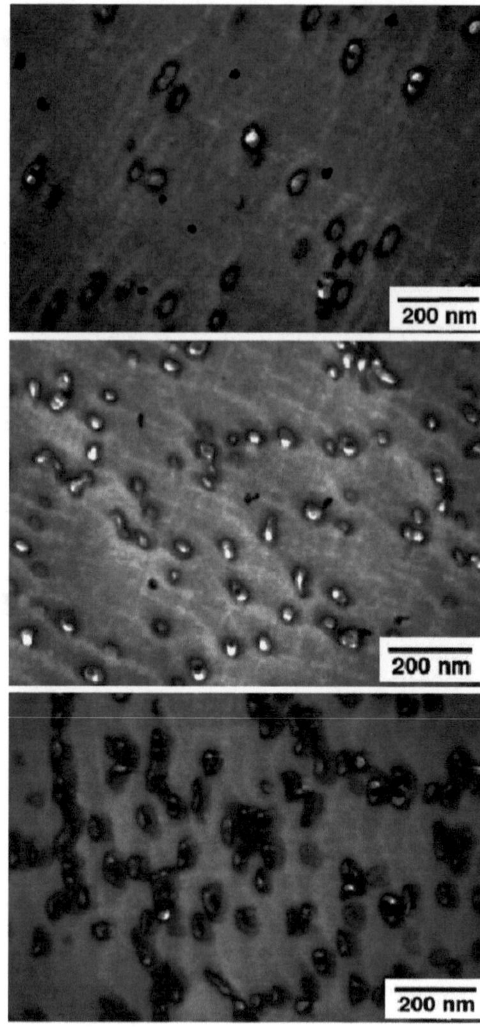

Fig. 35 TEM micrographs of the 75/25 PPE/PS blend modified with 5 wt% (*top*), 10 wt% (*middle*), and 20 wt% (*bottom*) of SIBS [65]. Published by permission of the Society of Chemical Industry

aging of the PPE film. Incorporation of charge trapping sites by compounding the PPE/PS blend with the third component SIBS also leads to an improvement of the charge storage properties.

5 Polyetherimide Electrets

As mentioned in the introduction, the commercial polyetherimide Ultem®1000 reveals the most promising electret behavior among the other high performance polymers investigated. However, a chemically identical polyetherimide synthesized

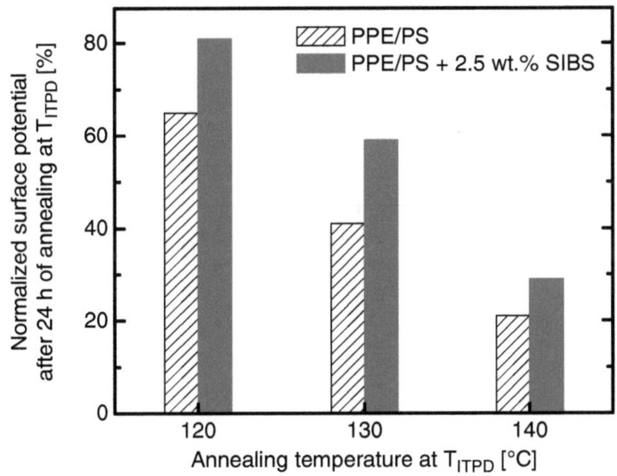

Fig. 36 Surface potential retained for the 75/25 PPE/PS blend films modified with 2.5 wt% SIBS after annealing for 24 h at a T_{ITPD} of 120, 130, and 140 °C [65]. Published by permission of the Society of Chemical Industry

in our laboratory showed an extremely poor charge storage capability (Fig. 37) [68]. Both samples were compression molded in the very same manner and subjected to the same ITPD experiments. The commercial Ultem® 1000 sample retained 75% of the initial surface potential, whereas the synthesized PEI **7a** loses its surface potential rapidly, resulting in an almost complete charge drain-off within 24 h. The observed discrepancy in the electret performance can be attributed to additives incorporated in the Ultem® 1000 PEI which act as electret-enhancing agents in PEI.

In order to corroborate the above assumption we identified additives present in the commercial material and subsequently compounded these additives into purified PEI (denoted as $\mathrm{PEI}_{\mathrm{pur}}$) to prove their function as electret-enhancing additives. We also investigated the influence of water and the influence of physical aging on the electret performance of PEI electret films. In addition, another approach to improve the charge storage capability of PEI by chemically modifying the polymer itself will be presented.

5.1 Additives in Polyetherimide Ultem® 1000

Purification of commercial Ultem® 1000 by reprecipitation and extraction will remove low-molecular-weight compounds, such as oligomers and additives, typically added as processing aids, plasticizers, antistatics, and most importantly for high temperature polymers, antioxidants. It was found that with each purification step the charge storage capability of Ultem® 1000 deteriorated dramatically, which we attributed to the removal of additives acting as charge traps [68]. The presence of

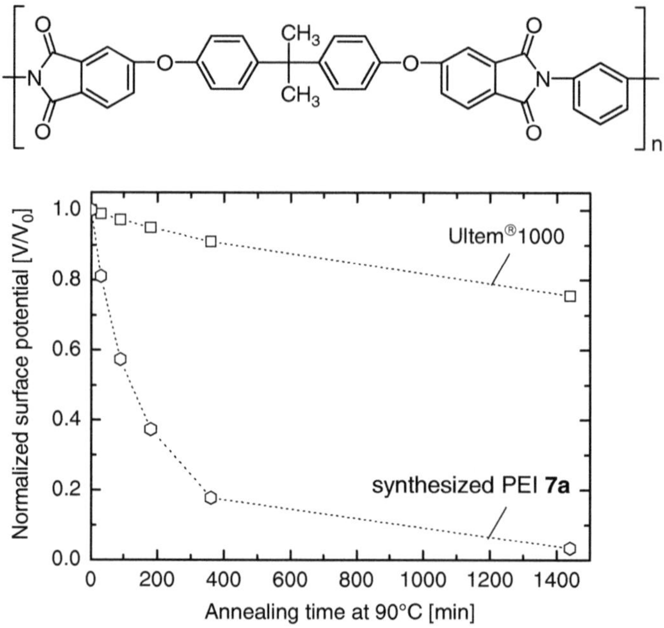

Fig. 37 Chemical structure of PEI (identical for Ultem®1000 and **7a**) and isothermal surface potential decay (ITPD) curves of compression molded and corona charged films of Ultem®1000 and **7a** [68]. Published by permission of Wiley Periodicals Inc

charge traps resulting from additives in commercial Ultem®1000 would explain the better ITPD performance of the as received samples compared to the purified samples. It would be beneficial in the scope of this work to identify chemically the extracted compounds, since they might function as electret additives in a similar manner as the additives for i-PP discussed in Sect. 3.

To identify the postulated charge trap additives in Ultem®1000, the residue in the extract obtained after the precipitation procedure was purified and characterized by spectroscopic methods. Besides PEI oligomers, one majority compound was present which was identified as an organophosphonite identical to the antioxidant sold under the trade names Irgafos®P-EPQ (Ciba) or Sandostab®P-EPQ (Clariant). The chemical structure is depicted in Fig. 38 [68]. Secondary antioxidants – in this case bulky organophosphorus compounds – are commonly employed as high temperature heat stabilizers. These P(III) compounds prevent degradation and oxidation of the molten polymer by hydroperoxides and P(V) species are formed during processing [69–71].

Next Irgafos®P-EPQ had to be confirmed in its assumed function to improve the electret performance of PEI. Therefore Irgafos®P-EPQ was melt compounded with purified PEI_{pur} in a twin-screw mixer at 366 °C and batches with different composition were prepared. In a first series, an initial additive concentration of 5,000 ppm (0.5 wt%) Irgafos®P-EPQ was chosen, which was further diluted by adding PEI_{pur},

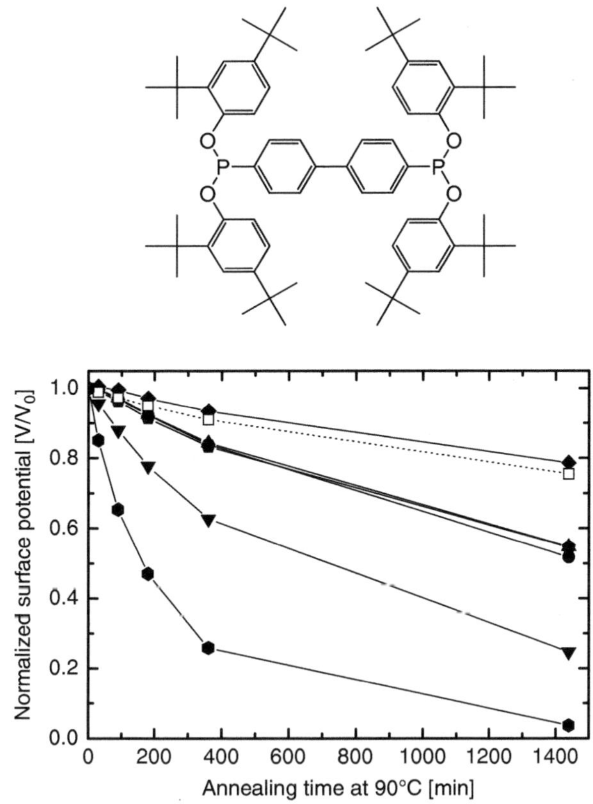

Fig. 38 Chemical structure of Irgafos®P-EPQ (*top*) and isothermal surface potential decay (ITPD, *bottom*) curves of melt compounded, compression molded, and corona charged films of PEI_{pur} additivated with 5,000 (*filled diamonds*), 3,200 (*filled pentagons*), 2,200 (*filled triangles*), 1,200 (*filled circles*), 700 (*filled inverted triangles*), and 0 ppm (*filled hexagons*) Irgafos®P-EPQ. For comparison, the curve for commercial Ultem®1000 films (*open squares*) is also included [68]. Published by permission of Wiley Periodicals Inc

yielding additive load levels of 3,200, 2,200, 1,200, and 700 ppm. PEI_{pur} was also extruded under the same processing conditions to supply a control sample. All films were obtained by compression molding at 320 °C, were corona charged, and their electret performance was determined by ITPD measurements.

Figure 38 shows the ITPD curves of PEI_{pur} films with Irgafos®P-EPQ in comparison to a commercial Ultem®1000 film. In the sample without additives nearly all surface potential is depleted within 24 h at 90 °C. Adding Irgafos®P-EPQ clearly improves the charge storage capability of PEI_{pur}. A pronounced improvement of the electret performance with increasing additive level was observed. Films with the highest concentration of 5,000 ppm Irgafos®P-EPQ exhibited a retained surface potential of 79%, which is slightly better than that of commercial Ultem®1000 films (75%). Films with lower additive concentrations of 3,200 ppm, 2,200 ppm, and 1,200 ppm show decay curves with very similar surface potentials of

52–55%. These results suggest that, in this concentration range, the ratio of oxidized P(V) and phosphonite P(III) species is balanced. Hence, the electret performance is dominated by the organophosphite Irgafos®P-EPQ acting as charge trap and not by thermal degradation products. At the lowest concentration of 700 ppm the film exhibited a charge retention of only 25%, which is still distinctly better than that of the purified PEI_{pur} (3%).

The mechanism of Irgafos®P-EPQ as secondary antioxidant is based on the oxidation of phosphonite P(III) to phosphonate P(V). Based on the dramatic charge drain of thermally stressed PEI samples, we conclude that only the added phosphonite P(III) species is capable of acting as a charge trap in PEI and not the oxidized P(V) derivative. Consequently, the added Irgafos®P-EPQ acts in two ways, suppressing thermal degradation, which itself accelerates charge decay, and as a charge trap.

To assess the scope of commercial secondary antioxidants as electret enhancers, we selected Irgafos®126 as another P(III) compound as additive for PEI. Irgafos®126 [72, 73], (Fig. 39) is a known organophosphite which is oxidized during heat stabilization, forming organophosphates. Irgafos®126 was melt compounded with PEI_{pur} in the same fashion as described above for Irgafos® P-EPQ. Electret films with concentrations of 5,000, 3,200, and 1,200 ppm were corona charged and subjected to ITPD measurements. Films of PEI_{pur} containing

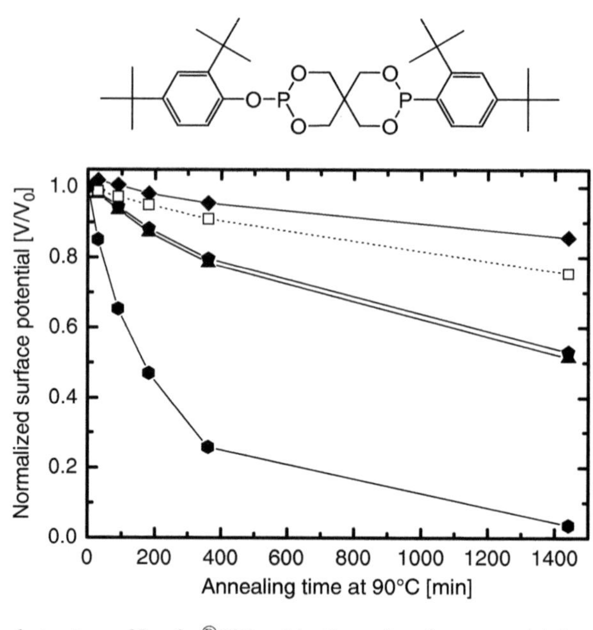

Fig. 39 Chemical structure of Irgafos® 126 and isothermal surface potential decay (ITPD) curves of melt compounded, compression molded, and corona charged films of PEI_{pur} additivated with 5,000 (*filled diamonds*), 3,200 (*filled pentagons*),1,200 (*filled triangles*), and 0 ppm (*filled hexagons*) Irgafos® 126. For comparison, the ITPD curve of commercial Ultem® 1000 films (*open squares*) is also included [68]. Published by permission of Wiley Periodicals Inc

5,000 ppm Irgafos®126 exhibited a remaining surface potential of 86% and hence superior to the electret performance of the commercial Ultem®1000 (75%). For films with concentrations of 3,200 ppm and 1,200 ppm Irgafos®126 in PEI_{pur}, ITPD curves comparable with curves obtained with compounds containing Irgafos®P-EPQ were measured, with retained charges of 50% after 24 h.

In summary, at comparable concentrations, films containing Irgafos®126 are superior electret materials compared to films containing Irgafos®P-EPQ. This result is in agreement with the fact that Irgafos®126 is a more efficient antioxidant than Irgafos®P-EPQ [70].

5.2 *Influence of Water Content*

In order to elucidate the effect of water absorption on the charge storage behavior of commercial Ultem®1000 films (50 μm thickness; Lipp-Terler), samples with various water contents were prepared. The water content was determined by a weight gain method using a Mettler Toledo MX5 micro-balance with a resolution of 1 μg. The water content of as-received films was determined by Karl-Fischer analysis to 0.74 wt%, compared to an equilibrium water absorbance of PEI at 1.39 wt% [27, 74]. As a preconditioning step, the specimens were dried under vacuum at 125 °C for 24 h to ensure a water content of less than 0.02%. By additional immersion treatments, the water content of these PEI films was increased to films with 0.45 wt% and 0.66 wt%.

Figure 40 compares the ITPD curves of Ultem®1000 films containing different amounts of water. The stability of the surface potential decreases with increasing amounts of water in the Ultem®1000 films. The largest decay in surface potential

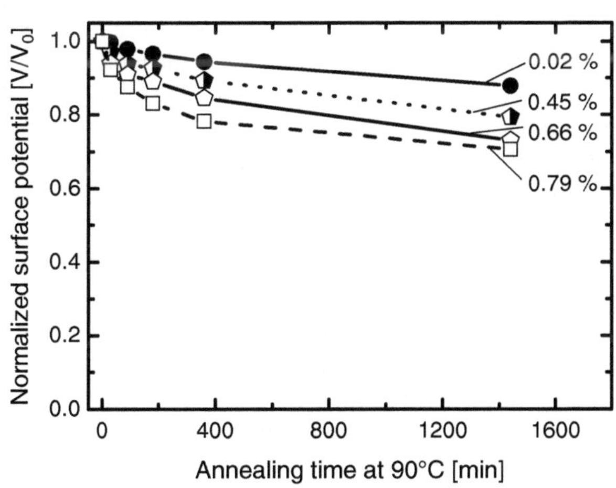

Fig. 40 Isothermal surface potential decay (ITPD) curves of Ultem®1000 films (50 μm thickness) with 0.02, 0.45, 0.66, and 0.79 wt% absorbed water as a function of the annealing time at 90 °C [27]. Published by permission of WILEY-VCH

was observed for the highest water content of 0.79 wt%, which is identical to the amount of water in commercial as-received films. After the annealing period of 24 h at 90 °C, 71% of the potential was still present. In conclusion, these results demonstrate that water in PEI films has a moderate influence on the electret behavior but does not cause a dramatic potential loss.

Thermal degradation products formed during melt processing may inflict the electret performance of polymers. To purify further the melt compressed films, individual compression molded films of commercial Ultem® 1000 were immersed in deionized water for 30 h at 90 °C, for 7 days at 80 °C, and for 4 days at 23 °C. To provide comparable conditions before charging, all soaked films were dried at 110 °C for 14 h without vacuum, and then equilibrated at ambient atmosphere and temperature for 1 h before corona charging. This treatment was necessary since on one hand it was practically impossible to charge films reproducibly without this equilibration step and on the other hand this much better reflects conditions encountered in possible applications. According to Karl-Fischer analysis, compression molded films of Ultem® 1000 contained 0.50 wt% and compression molded, immersed, and subsequently dried films of Ultem® 1000 contained 0.46 wt% water [68]. These almost identical values indicate that the films very quickly absorb water up to this value during the equilibrating step.

Figure 41 (top) depicts the ITPD curves. Surprisingly, all films stored in water exhibit a distinctly *better* charge storage capability than untreated Ultem® 1000 films. Films immersed in water at 80 °C for 7 days maintain almost their initial surface potential of 98%. A shorter immersion time in combination with a lower temperature, more precisely 4 days at 23 °C, renders electrets which retained 94% of the initial value. Also an excellent electret (97%) are films which were immersed in water for a shorter period of time, but at a higher temperature (30 h at 90 °C). To interpret these results, we argue that polar compounds and degradation products have the potential to accelerate the charge decay in electret materials. It is also feasible that, during the immersion in water, these polar compounds were removed or diminished by diffusion into the water phase, and this process is most efficient using hot water.

In addition to the excellent charge retention, the ITPD curves of films immersed in water at 80 °C and 90 °C (magnified view of Fig. 41, bottom) show a maximum after 90 min at 90 °C, indicated by a relative surface potential of 103% and 102%, respectively. This maximum is reproducible and not an artifact. Such an increase of the surface potential is documented in the literature [75, 76] and is due to the coexistence of hetero- and homocharges. Surface charges with the same polarity as the corona electrode are called homocharges and are real charges [1–3]. Heterocharges, however, are linked to dielectric absorption involving dipoles or ionic charges and are aligned opposite to the applied field during charging. The observed initial increase in net homocharge is due to the decay of the heterocharge and is just what one would expect if both heterocharges and homocharges were present in a net homocharge sample [75]. Generally, only in very stable corona charged polymer electrets containing heterocharges can this initial increase in the ITPD curve be observed. If heterocharges are absent, this effect is not observed.

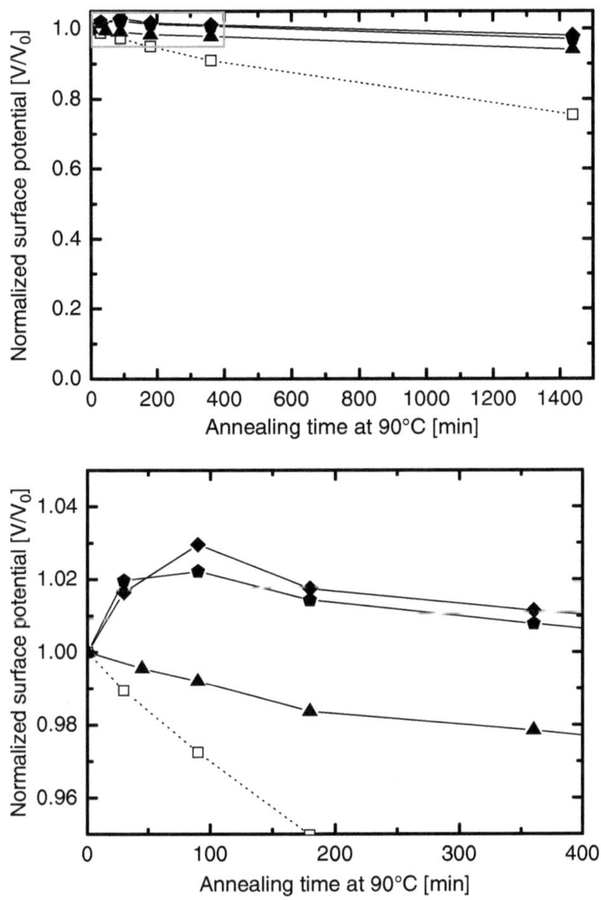

Fig. 41 Isothermal surface potential decay (ITPD) curves of compression molded (320 °C) films of Ultem® 1000 which were immersed in water at 90 °C for 30 h (*filled pentagons*), at 80 °C for 7 days (*filled diamonds*), and at 23 °C for 4 days (*filled triangles*), dried, and corona charged (*top*). For comparison, the curve for untreated compression molded Ultem® 1000 films (*open squares*) is also included. The *bottom graph* shows a magnified view of the *region boxed in the top graph* [68]. Published by permission of Wiley Periodicals Inc

5.3 *Physical Aging of Ultem® 1000 Films*

As summarized in Sect. 4.3, physical aging of amorphous polymers can improve the resulting electret performance. Consequently, now PEI is subjected to such a treatment and the impact on the electret properties is investigated. The degree of physical aging can easily be monitored by DSC; during the controlled heating of an aged sample, an endothermic peak reflecting the enthalpy recovery appears around T_g [77]. Hodge et al. described this phenomenon in great detail and reported DSC curves and the corresponding enthalpy relaxations (Δh) of various amorphous physically aged polymers [78–80].

Therefore non-aged Ultem®1000 films were compared to Ultem®1000 films aged at $T_{age} = 200\,°C$, which is below the T_g of PEI at 220 °C. The aging periods (t_{age}) varied between 30 min and 4 days. The enthalpy relaxation (Δh) was calculated from DSC traces according to Hutchinson by the area subtraction method [59, 81]. The enthalpy relaxations of PEI Ultem®1000 as a function of aging time are displayed in Fig. 42. The Δh value continuously increases with increasing physical aging time and levels off. Ultem®1000 reaches a low Δh during the starting period of aging, but over longer times reaches a Δh of 3.73 J/g. Belana et al. also studied the physical aging of Ultem®1000; however their maximum aging period was only up to 48 h [77]. Within this range their reported Δh values agree with our results.

The results for the ITPD curves at $T_{ITPD} = 120\,°C$ for Ultem®1000 are depicted in Fig. 43. Note the difference between the physical aging temperature (T_{age}), aging time (t_{age}), the charge decay acceleration temperature (T_{ITPD}), and time (t_{ITPD}), respectively. For the ITPD of the non-aged Ultem®1000, a pronounced charge drain-off is observed. After 24 h at $T_{ITPD} = 120\,°C$ only 18% of the initially applied surface potential is still present. When Ultem®1000 was aged at 200 °C for 30 min the charge decay is slightly delayed. An elongation of the aging process to 24 h raises the charge storage capability significantly, finally reaching its maximum at 95% when the film was aged for 4 days at 200 °C. Although this aging process is energy and time consuming, Ultem®1000 can be converted to an electret material with outstanding high temperature stability with 95% of the initially applied charge retained after 24 h at $T_{ITPD} = 120\,°C$.

Figure 44 presents the TSC spectra of non-aged and aged compression molded and charged films of Ultem®1000. As mentioned before, amplitude and location of

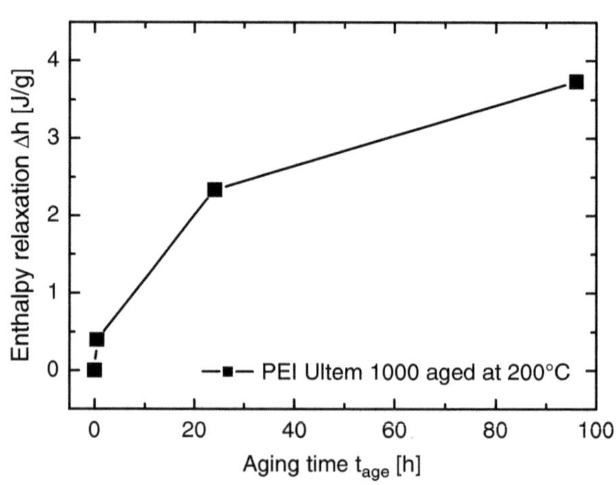

Fig. 42 Dependence of the enthalpy relaxation (Δh) at 20 K/min on the aging time of compression-molded Ultem®1000 films at an aging temperature T_{age} of 200 °C

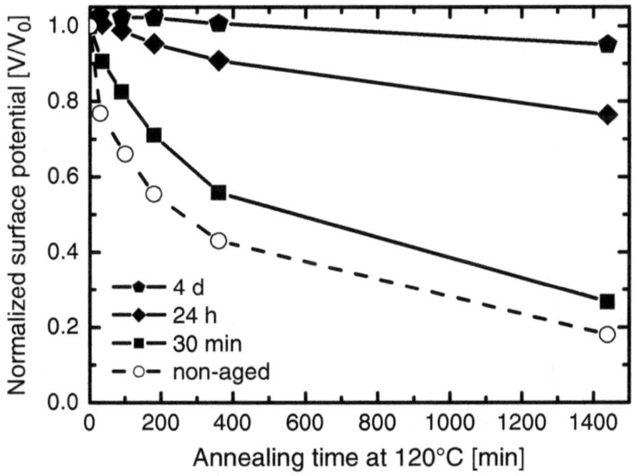

Fig. 43 Isothermal surface potential decay (ITPD) curves as a function of time at $T_{ITPD} = 120\,^{\circ}C$ of compression-molded Ultem®1000 films aged at $T_{age} = 200\,^{\circ}C$

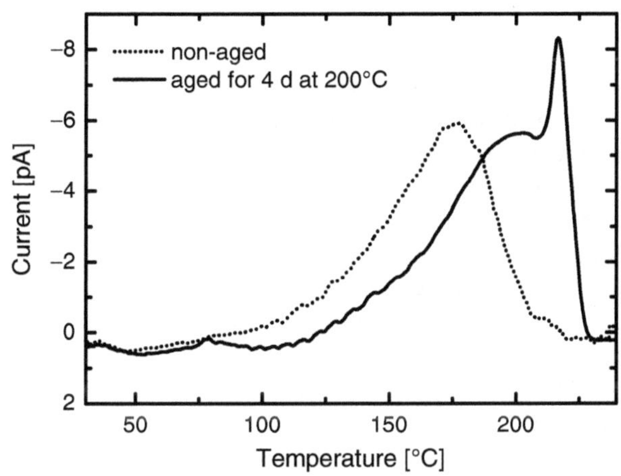

Fig. 44 Thermally stimulated discharge (TSC) curves recorded at a heating rate of 200 K/h of non-aged and aged compression-molded Ultem®1000 films. Films of 100 μm thickness were corona charged to +400 V for 20 s

the resulting peaks reflect the characteristics of the trapping depths for charges in the material. The non-aged Ultem®1000 film shows a very broad peak between 100 °C and 220 °C with a maximum around 175 °C. According to the literature, three sub-relaxations are located in this temperature range, denoted as β relaxations [82, 83]. In aged Ultem®1000, this peak is shifted to a higher temperature value with a maximum at 200 °C. Compared to the non-aged Ultem®1000, the broad peak has been shifted both in the onset and in the maximum by around 50 °C. Moreover, in the aged Ultem®1000 film a new peak at 220 °C has developed which corresponds clearly to

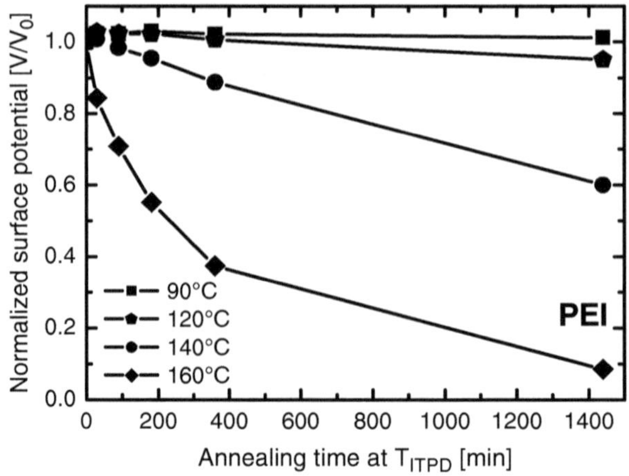

Fig. 45 Isothermal potential decay (ITPD) as a function of the time at T_{ITPD} of compression-molded and corona charged Ultem®1000 films, aged at $T_{\mathrm{age}} = 200\,^{\circ}\mathrm{C}$ for $t_{\mathrm{age}} = 4$ days prior to corona charging

the T_g of the polymer. This peak is not visible in the non-aged Ultem®1000 since the films are completely discharged within the β relaxation before 220 °C is reached.

TSC spectra hold information about the temperature above which the charge decay progresses, indicated by the onset of the discharge. However, for many applications the charge decay at temperatures *below* the first discharge peak is of great interest. Therefore ITPD measurements of aged Ultem®1000 at elevated temperatures were performed since these measurements are particularly interesting for the development of electret materials with high temperature stability.

Figure 45 depicts the ITPD curves of aged Ultem®1000 films ($T_{\mathrm{age}} = 200\,^{\circ}\mathrm{C}$ for $t_{\mathrm{age}} = 4$ days), which were evaluated at $T_{\mathrm{ITPD}} = 90\,^{\circ}\mathrm{C}$, 120 °C, 140 °C, and 160 °C. The aged Ultem®1000 films were capable of retaining the entire initially applied surface charge after 24 h at $T_{\mathrm{ITPD}} = 90\,^{\circ}\mathrm{C}$ and lost only 5% at 120 °C. Above this temperature ($T_{\mathrm{ITPD}} = 140\,^{\circ}\mathrm{C}$ and 160 °C), where the onset of the β peak at 125 °C is reached (see Fig. 44), the charges drain off faster. After 24 h at $T_{\mathrm{ITPD}} = 140\,^{\circ}\mathrm{C}$ the aged Ultem®1000 sample still maintains 60% of its initially applied charge, which is three times better than the non-aged Ultem®1000 at $T_{\mathrm{ITPD}} = 120\,^{\circ}\mathrm{C}$ shown in Fig. 43.

5.4 Fluorinated Polyetherimides as Electret Materials

Another approach to improve the charge storage capability of polymers is the synthesis of specially tailored materials. The structure variations and the synthesis conditions of the polymers have to be optimized to comply with the corona charging method and the final application. Suitable polymers have to be of low-k type, since a high dielectric constant leads to dipole orientation during charging and thus neutralizes the trapped homocharges. Further, the T_g of the polymer should be considerably higher than the application or ITPD temperature.

It is well known that the introduction of fluorine atoms in the polymer backbone can efficiently reduce the dielectric constant [1]. In addition, fluorinated moieties usually improve the charge storage capability. Based on this knowledge, we modified the polymer structure of PEI by introducing trifluoromethyl groups and thus obtained polymers containing increasing amounts of fluorine [84]. Scheme 4 displays the chemical structure of the two synthesized series of fluorinated polyetherimides (F-PEI) based on 4,4′-(4,4′-isopropylidenediphenoxy) bis(phthalic anhydride) **7** and 4,4′-(4,4′-hexafluoroisopropylidenediphenoxy) bis(phthalic anhydride) **8**. PEI **7a** will again serve as a reference polymer, since it contains no fluorine and is structurally identical to the commercial PEI Ultem®1000, but is free of any additives which might alter the electret performance.

Table 3 summarizes the fluorine content, inherent viscosities, and thermal properties of the synthesized fluorinated polyetherimides The inherent viscosities (for conditions, see footnote of Table 3) of the F-PEI polymers varied from 0.34 to 0.72 dL/g. PEI **7a** was obtained at an inherent viscosity of 0.47 dL/g, which is comparable to the value of commercial Ultem®1000 (η_{inh} =0.5 dL/g). DSC measurements showed that all F-PEIs are amorphous with a T_g range from 220 °C to 247 °C. Within the first series denoted as **7a**, **7b**, and **7d**, T_g between 220 and 224 °C were detected, whereas **7c** exhibits a slightly higher T_g of 232 °C. Polymers **8a**, **8b**, and **8d** show T_g between 234 and 236 °C, whereas the T_g of **8c** is at 247 °C.

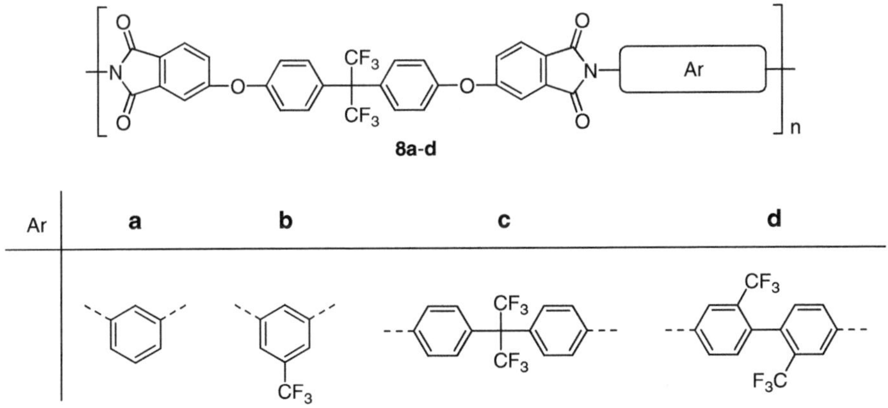

Scheme 4 Chemical structures of the synthesized fluorinated polyetherimides

Table 3 Fluorine content, inherent viscosities, and thermal properties of FPEIs **7a–d** and **8a–d**

Polymer	Fluorine content [wt%]	η_{inh}[a] [dL/g]	T_g[b] [°C]	$T_{5\%}$[c] [°C]
Ultem®1000[d]	0	0.50	221	535
7a	0	0.47	221	535
7b	8.6	0.34	220	530
7c	13.9	0.50	232	530
7d	14.2	0.46	224	495
8a	16.3	0.66	234	535
8b	22.2	0.40	236	540
8c	24.6	0.72	247	530
8d	25.0	0.53	235	535

[a]Inherent viscosity of a 0.5 g/dL solution in NMP at 25 °C
[b]DSC: heating rate 10 K/min; data of the second heating run
[c]TGA: heating rate 10 K/min in air; temperature at 5 wt% loss
[d]Commercial pellets obtained from GE plastics

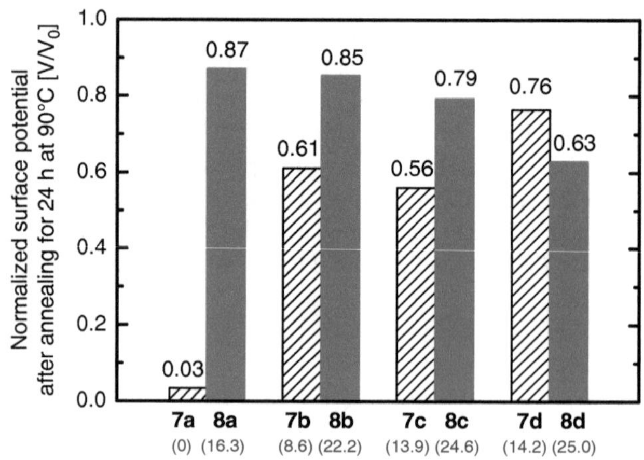

Fig. 46 Remaining surface potential of fluorinated PEIs after 24 h at 90 °C. The fluorine content in wt% of each polymer is given *in parentheses below*. The compression molded films of 100 μm thickness were corona charged to +400 V for 20 s [84]. Published by permission of WILEY-VCH

Comparing each structure of the first series **7a–d** to the structurally similar polymers of the second series **8a–d**, the second series possesses T_g which are 11–16 °C higher. This indicates that by replacing two methyl groups with two trifluoromethyl groups in the dianhydride moiety, the flexibility of the polymer chain decreases. All synthesized F-PEIs exhibit excellent temperature stability: In air they show a 5 wt% loss between 530 °C and 540 °C, but for **7d**, the $T_{5\%}$ is at 495 °C.

Compression molded F-PEI films of 100 μm thickness were utilized to investigate their charge storage characteristics. After corona charging to +400 V for 20 s, ITPD measurements at 90 °C were conducted. Figure 46 compares the remaining normalized surface potentials of all F-PEIs after 24 h at 90 °C.

In the case of the non-fluorinated PEI **7a** synthesized with *m*-phenylene diamine as comonomer, a fast charge decay occurs and a residual surface potential of only 3% remained. By exchanging the dianhydride moiety **7** with its hexafluorinated analog, F-PEI **8a** was obtained which is an outstanding electret material: The renaming surface potential after 24 h at 90 °C is as high as 87%. Introducing an additional trifluoromethyl group into the diamine comonomer by using 3,5-diaminobenzotrifluoride yields the F-PEIs **7b** and **8b**. Again, **8b** based on the hexafluorinated dianhydride outperforms **7b** with residual surface potentials of 85% and 61%, respectively. The charge retention of **8b** is at a similarly high level as **8a**. However, the electret performance of **7b** is clearly improved compared to **7a**. These results indicate that the introduction of only one trifluoromethyl group into the diamine moiety of the polymer improves the electret performance significantly. Very similar is the behavior of **7c** (56%) compared to **8c** (79%). Here the polymer with the higher fluorine content (**8c**) exhibits the better charge retention. Peculiarly, the polyetherimides composed with the non-coplanar hexafluorinated diamine, **7d** and **8d**, show a different behavior. Here, **8d** exhibits a better electret performance with charge retention of 76% compared to **8d** with 63%.

From a general point of view, due to the amorphous nature of PEI a large number of localized states ("traps") are present, which are formed by dipole–dipole interactions of the polymer chains, resulting in internal fields which delay the charge drift in the bulk film. The introduction of fluorine modifies the molecular electronics levels due to the high electronegativity of the fluorine atoms. It is remarkable that by replacing only two methyl groups with trifluoromethyl groups in the dianhydride moiety of the PEI, the charge retention capability in the electret films is improved from poor to very good.

6 Conclusions and Outlook

In this chapter we have presented concepts and methods to improve polymer electret materials. *Corona charging* was employed for film charging since this method possesses the advantages of being cost-effective, easy to perform, and applicable in short periods of time.

The commodity polymer *isotactic polypropylene* (i-PP) is known to exhibit good charge storage capability with moderate surface potential decays at temperatures up to 90 °C. We demonstrated that certain additives based on substituted 1,3,5-benzene trisamides and bisamides are efficient *additives* to improve the electret properties of i-PP. Due to their chemical structure which enables H-bonds, these additives self-assemble in the i-PP matrix and these supramolecular aggregates act as charge traps, effectively hindering the charge drift. We found for an optimization of this effect, the additive concentration has to be decreased to a very low level. Further improvement of the charge storage behavior could be achieved by increasing the cooling rate applied during film processing. These results obtained using selected compounds clearly demonstrate the potential of this class of additives to provide optimal i-PP electret materials.

However, the range of application of i-PP is limited by its continuous service temperature of 70 °C. This issue inevitably requires other polymer systems which withstand *higher temperatures*. On this account, we presented experimental results highlighting the potential of neat *poly(2,6-dimethyl-1,4-phenylene ether)* (PPE) and of *poly(2,6-dimethyl-1,4-phenylene ether)/polystyrene* (PPE/PS) blends as materials possessing an excellent electret performance even at elevated temperatures. The oxidative degradation of neat PPE during melt processing constitutes a drawback that can be overcome by blending with polystyrene. Particularly, PPE/PS blend with a composition of 75/25 exhibited excellent charge storage capability, even at temperatures up to 120 °C. Incorporation of charge trapping sites by compounding the PPE/PS blend with the *third component SIBS* also leads to an improvement of the charge storage properties. The aforementioned results demonstrate that *blending* is a promising approach towards satisfying the demands of electret applications.

With the commercial *polyetherimide* (PEI) Ultem® 1000 we presented another high performance polymer which possesses promising charge storage characteristics. We successfully removed and identified phosphorus-based *antioxidant additives*, which are responsible for the good electret performance. Reincorporation of such additives into purified PEI demonstrated the beneficial effect. However, outstanding electrets can also be achieved by modifying the chemical structure of PEI. Introducing trifluoromethyl groups at the right position in polyetherimides distinctly improves the charge storage capabilities in comparison to non-fluorinated PEI.

A very simple but yet time-consuming treatment to improve amorphous electret materials is *physical aging* prior to corona charging. For instance, the charge retention of Ultem® 1000 films after 24 h at 120 °C can be raised from 18% to 95% when the films were physically aged at 200 °C for 4 days.

The materials and methods presented in this review allow access to charge storage materials with extraordinary high temperature stability which can be employed for instance in microphones, sensor devices, and electret filters. In addition, such excellent electrets are of fundamental importance for *piezoelectric devices* within the rapidly emerging field of energy harvesting.

Acknowledgments The authors are deeply indebted to Dr. Nils Mohmeyer and Dr. Nico Behrendt for their dedicated work during their Ph.D. theses and to Dr. Thomas Frese for his contributions during his post-doctoral visit at the University of Bayreuth.

Special thanks go to Prof. Gerhard M. Sessler and to Dr. Joachim Hillenbrand (Technical University of Darmstadt) for their continuous support and many fruitful discussions on the physics and characterization of our polymer electret materials. Additional thanks for performing the TSD measurements.

Jutta Failner, Sandra Ganzleben, and Doris Hanft (Makromolekulare Chemie I, University of Bayreuth) are gratefully acknowledged for their invaluable contributions in material synthesis and sample preparation.

The authors are grateful to the German Science Foundation (DFG) for generously providing financial support within the Collaborative Research Center (SFB) 481, project B11. The part on polypropylene was funded within the DFG program "Vom Molekül zum Material" (Schm 703/2–2; Al 474/5–2).

References

1. Kao KC (2004) Dielectric phenomena in solids. Elsevier Academic Press, San Diego
2. Sessler GM (1998) Electrets, vol 1. Laplacian Press, Morgan Hill
3. Hilczer B, Malecki J (1986) Studies in electrical and electronic engineering. Elsevier Science Publishers BV, Amsterdam
4. Giacometti JA, Fedosov S, Costa MM (1999) Brazil J Phys 29:269–279
5. Heaviside O (1885) The Electrician 230. Republished in: Heaviside O (1892) Electrical Papers 1:488–493
6. Eguchi M (1919) Proc Phys Math Soc Jpn 1:320–326
7. Eguchi M (1919) Proc Phys Math Soc Jpn 1:326–331
8. Gross B (1944) Phys Rev 66:26–28
9. Campo EA (2008) Selection of polymeric materials: how to select design properties from different standards. William Andrew Inc, Norwich
10. Gerhard-Multhaupt R, Gross B, Sessler GM (1998) Recent progress in electret research. In: Sessler GM (ed) Electrets, vol 1. Laplacian Press, Morgan Hill
11. Mascarenhas S (1998) Bioelectrets: electrets in biomaterials and biopolymers. In: Sessler GM (ed) Electrets, vol. 1. Laplacian Press, Morgan Hill
12. Klimiec E, Zaraska W, Zaraska K, Gasiorski KP, Sadowski T, Pajda M (2008) Microelect Reliab 48:897–901
13. Shen D, Choe S-Y, Kim D-J (2007) Jpn J Appl Phys Part 1 46:6755–6760
14. Farinholt KM, Pedrazas NA, Schuluneker DM, Burt DW, Farrar CRJ (2009) Intel Mater Syst Struct 20:633–642
15. Cook-Chennault KA, Thambi N, Sastry AM (2008) Smart Mater Struct 17: 043001/043001–043001/043033
16. Anton SR, Sodano HA (2007) Smart Mater Struct 16: R1–R21
17. Hudak NS, Amatucci GG (2008) J Appl Phys 103:101301/101301–101301/101324
18. Mishra A (1982) J Appl Polym Sci 27:381–395
19. Yagishita A, Yamanouchi H, Ikezaki K (1999) Jpn J Appl Phys Part 1 38:2053–2058
20. Arita Y, Sha Shiratori S, Ikezaki K (2003) J Electrost 57:263–271
21. Mishra A (1982) J Appl Polym Sci 27:1107–1118
22. Perlman MM (1972) J Electrochem Soc 119:892–898
23. Wintle HJ (1973) J Acoust Soc Am 53:1578–1588
24. Wintle HJ (1999) IEEE Trans Dielectr Electr Insul 6:1–10
25. von Salis-Soglio C (2010) Improving the electret and electrospinning performance of isotactic polypropylene by additives. Doctoral thesis, University of Bayreuth, Bayreuth, Germany
26. Zweifel H, Maier RD, Schiller M (2009) Plastics additives handbook, 6th edn. Carl-Hanser-Verlag, München
27. Frese T, Lovera D, Sandler JKW, Lim GT, Altstädt V, Giesa R, Schmidt H-W (2007) Macromol Mater Eng 292:582–587
28. Platt D (2003) Engineering and high performance plastics. Rapra Technology Ltd, Shawbury, pp 9–11
29. Mohmeyer N, Müller B, Behrendt N, Hillenbrand J, Klaiber M, Zhang X, Smith P, Altstädt V, Sessler GM, Schmidt H-W (2004) Polymer 45:6655–6663
30. Mohmeyer N, Schmidt H-W, Kristiansen PM, Altstädt V (2006) Macromolecules 39: 5760–5767
31. Mohmeyer N, Behrendt N, Zhang X, Smith P, Altstädt V, Sessler GM, Schmidt H-W (2007) Polymer 48:1612–1619
32. Behrendt N, Mohmeyer N, Hillenbrand J, Klaiber M, Zhang X, Sessler GM, Schmidt H-W, Altstädt V (2006) J Appl Polym Sci 99:650–658
33. Behrendt N, Greiner C, Fischer F, Frese T, Altstädt V, Schmidt H-W, Giesa R, Hillenbrand J, Sessler GM (2006) Appl Phys A Mater Sci Process 85:87–93
34. van Turnhout J (1998) Thermally stimulated discharge of electrets. In: Sessler GM (ed) Electrets, vol. 1. Laplacian Press, Morgan Hill

35. Mittal A, Jain V, Mittal JJ (2001) Mater Sci Lett 20:681–685
36. Hillenbrand J, Behrendt N, Altstädt V, Schmidt H-W, Sessler GM (2006) J Phys D Appl Phys 39:535–540
37. Nago S, Mizutani Y (1998) J Appl Polym Sci 68:1543–1553
38. Blomenhofer M, Ganzleben S, Hanft D, Schmidt H-W, Kristiansen M, Smith P, Stoll K, Mäder D, Hoffmann K (2005) Macromolecules 38:3688–3695
39. Kristiansen PM, Gress A, Smith P, Hanft D, Schmidt H-W (2006) Polymer 47:249–253
40. Schmidt H-W (2007) Proceedings of the 5th European Conference on "Additives and Colors", Mondorf-les-Bains, Luxembourg, Mar. 14–15, 2007, pp 124–132
41. Hillenbrand J, Motz T, Sessler GM, Zhang X, Behrendt N, von Salis-Soglio C, Erhard DP, Altstädt V, Schmidt H-W (2009) J Phys D Appl Phys 42:065410 (8pp)
42. Smitha B, Sridhar S, Khan AA (2003) J Membrane Sci 225:63–76
43. Kalkar AK, Kundagol S, Chand S, Chandra S (1996) J Appl Polym Sci 61:1431–1437
44. Hay AS (1998) J Polym Sci Part A Polym Chem 36:505–517
45. Lovera D, Ruckdäschel H, Göldel A, Behrendt N, Frese T, Sandler JKW, Altstädt V, Giesa R, Schmidt H-W (2007) Eur Polym J 43:1195–1201
46. Kalkar AK, Kundagol S, Chand S, Chandra S (1996) J Appl Polym Sci 61:1431–1437
47. Seike Y, Okude Y, Iwakura I, Chiba I, Ikeno T, Yamada T (2003) Macrom Chem Phys 204:1876–1881
48. Brandrup J, Immergut EH (1989) Polymer handbook, 3rd edn. Wiley, New York
49. Schwödiauer R, Neugschwandtner GS, Bauer-Gogonea S, Bauer S, Wirges W (1999) Appl Phys Lett 75:3998–4000
50. Shur YJ, Ranby B (1975) J Appl Polym Sci 19:1337–1346
51. Feng HQ, Feng ZL, Ruan HZ, Shen LF (1992) Macromolecules 25:5981–5985
52. Robertson CG, Wilkes GL (2000) Polymer 41:9191–9204
53. Fox TG (1956) Bull Am Phys Soc 1:123
54. Couchman PR (1978) Macromolecules 11:1156–1161
55. Wetton RE, Macknight WJ, Fried JR, Karasz FR (1978) Macromolecules 11:158–165
56. Macknight WJ, Karasz FE, Fried JR (1978) Solid state transition behavior in blends. In: Paul DR, Newman S (eds) Polymer blends, vol 1. Academic Press Inc, San Diego, pp 185–242
57. Li S, Dickinson LC, Chien JCW (1991) J Appl Polym Sci 43:1111–1116
58. Struik LCE (1977) Polym Eng Sci 17:165–173
59. Hutchinson JM (1995) Prog Polym Sci 20:703–760
60. Sperling LH (2006) Introduction to physical polymer science. Wiley-Interscience, New Jersey
61. Cowie JMG, Arrighi V (2008) Polymers: chemistry and physics of modern materials. CRC Press, Boca Raton
62. Liao K, Quan D, Lu Z (2001) Eur Polym J 38:157–162
63. Etienne S, David L (2007) Philos Mag 87:417–424
64. Huang Y, Paul DR (2004) Polymer 45:8377–8393
65. Lovera D, Ruckdäschel H, Schmidt H-W, Altstädt V (2009) Polym Int 58:1260–1266
66. Puskas JE, Kwon Y, Altstädt V, Kontopoulou M (2007) Polymer 48:590–597
67. Datta S, Lohse DJ (1996) Polymeric compatibilizers. Hanser, New York
68. Erhard DP, Giesa R, Altstädt V, Schmidt H-W (2010) J Appl Polym Sci 115:1247–1255
69. Ciba data sheet (1999) Tetrakis (2,4-di-*tert*-butylphenyl)[1,1-biphenyl]-4,4-diylbisphosphonite), CAS number 119345–01–6
70. Costanzi S, Farris R, Girelli D (2001) Polym Degrad Stab 73:425–430
71. Stevenson DR, Jakupca M, Farber J (2005) Proceedings of Addcon World International Conference, Sept. 21–22, 2005, 11th edn. Hamburg, Germany. Paper 18/11–18/21
72. Ciba data sheet (2000) Bis-(2,4-di-*tert*-butylphenol)pentaerythritol diphosphite, CAS number 26741–53–7
73. Stein D, Stevenson DJ (2000) Vinyl Addit Technol 6:129–137
74. Thominette F, Merdas I, Verdu J (2003) Humid ageing of polyetherimide: chemical and physical interactions with water. In: Mittal KL (ed) Polyimides and other high temperature polymers, vol 2. VSP, Utrecht
75. Cross JD (1968) J Electrochem Soc 115:42–45

76. Perlman MM, Reedyk CW (1968) J Electrochem Soc 115:45–49
77. Belana J, Canadas JC, Diego JA, Mudarra M, Diaz R, Friederichs S, Jaimes C, Sanchis M (1998) J Polym Int 46:29–32
78. Hodge IM, Huvard GS (1983) Macromolecules 16:371–375
79. Hodge IM (1983) Macromolecules 16:898–902
80. Hodge IM (1983) Macromolecules 20:2897–2908
81. Hutchinson JM, Smith S, Horne B, Gourlay GM (1999) Macromolecules 32:5046–5061
82. Sanchis MJ, Diaz-Calleja R, Jaimes C, Belana J, Canadas JC, Diego JA, Mudarra M, Sellares J (2004) Polym Int 53:1368–1377
83. Belana J, Canadas JC, Diego JA, Mudarra M, Diaz-Calleja R, Friederichs S, Jaimes C, Sanchis MJ (1998) Polym Int 46:11–19
84. Erhard DP, Giesa R, Altstädt V, Schmidt H-W (2007) Macromol Chem Phys 208:1522–1529

Index

MIX
Papier aus verantwortungsvollen Quellen
Paper from responsible sources
FSC® C105338

If you have any concerns about our products,
you can contact us on
ProductSafety@springernature.com

In case Publisher is established outside the EU,
the EU authorized representative is:
Springer Nature Customer Service Center GmbH
Europaplatz 3, 69115 Heidelberg, Germany

Printed by Libri Plureos GmbH
in Hamburg, Germany